Principles
Infrastruct

CW01271735

Principles of Project and Infrastructure Finance is an introductory book that treats project finance from a project manager's perspective. The approach is intuitive and yet rigorous, making the book highly readable. Case studies are used to illustrate integration as well as to underscore the pragmatic slant. The book is intended for graduate and senior undergraduate students, researchers, and parties involved in project finance. These include company directors, project managers, lenders, lawyers, architects, contractors, engineers, quantity surveyors, regulators, suppliers, insurers, and investors.

While there are many texts on each of the topics, and indeed on groups of such topics, there is no book that I have come across that takes this welcome holistic approach to project finance issues from a project manager's perspective.

Mohan Kumaraswamy
University of Hong Kong

This timely book provides an introduction to the key areas within project finance with relevant case studies in selected sectors. It is well-structured with an integrated map that guides and assists the busy project manager at one stop.

Jiang Hongbin
Asia Development Bank

Willie Tan is Associate Professor in the Department of Building and Vice Dean (Academic) at the School of Design and Environment, National University of Singapore. He is also the Program Director of the M.Sc. (Project Management) program, and Co-Director of the Center of Project Management and Construction Law.

Also available from Taylor & Francis

Handbook of Project Finance for Water and Wastewater Systems

M. Curley

Hb: ISBN 978–0–87371–486–0

Construction Economics

D. Myers

Hb: ISBN 978–0–415–28638–1
Pb: ISBN 978–0–415–28639–8

The Official History of Britain and the Channel Tunnel

T. Gourvish

Hb: ISBN 978–0–415–39183–2

Construction Project Management

P. Fewings

Hb: ISBN 978–0–415–35905–4
Pb: ISBN 978–0–415–35906–1

Construction Contracts 4th ed

W. Hughes et al.

Hb: ISBN 978–0–415–39368–3
Pb: ISBN 978–0–415–39369–0

Information and ordering details

For price availability and ordering visit our website **www.tandfbuiltenvironment.com**
Alternatively our books are available from all good bookshops.

Principles of Project and Infrastructure Finance

Willie Tan

Taylor & Francis
Taylor & Francis Group

LONDON AND NEW YORK

First published 2007
by Taylor & Francis
2 Park Square, Milton Park, Abingdon, Oxon OX14 4RN

Simultaneously published in the USA and Canada
by Taylor & Francis
270 Madison Ave, New York, NY 10016, USA

*Taylor & Francis is an imprint of the Taylor & Francis Group, an
informa business*

© 2007 Willie Tan

Typeset in Times New Roman by
Integra Software Services Pvt. Ltd, Pondicherry, India
Printed and bound in Great Britain by
Cpod, Trowbridge, Wiltshire

Publisher's Note
This book has been prepared from camera-ready copy provided by the
author.

British Library Cataloguing in Publication Data
A catalogue record for this book is available from the British Library

Library of Congress Cataloging in Publication Data
Tan, Willie.
Principles of project and infrastructure finance / Willie Tan.
p. cm.
Includes bibliographical references and index.
ISBN 978-0-415-41576-7 (hardback : alk. paper) --
ISBN 978-0-415-41577-4 (pbk. : alk. paper)
1. Capital investments. 2. Corporations--Finance. 3. Infrastructure
(Economics)--Finance. 4. Risk management. I. Title.
HG4028.C4T36 2007
658.15′5--dc22
2006039620

ISBN10: 0–415–41576–4 Hardback
ISBN10: 0–415–41577–2 Paperback
ISBN10: 0–203–96250–8 ebook

ISBN13: 978–0–415–41576–7 Hardback
ISBN13: 978–0–415–41577–4 Paperback
ISBN13: 978–0–203–96250–3 ebook

CONTENTS

Abbreviations and notations

The following abbreviations and symbols are used consistently in this book. Where symbols listed here are also used to denote something else, its meaning should be clear from the context.

IA	Implementation agreement
IPP	Independent power producer
IRR	Internal rate of return
JV	Joint venture
LIBOR	London interbank offered rate
NPV	Net present value
PPA	Power purchase agreement
SIBOR	Singapore interbank offered rate
SPV	Special purpose vehicle

C_0	initial cost
d	dividend
E_0	initial equity
F_t	cash flow for year t
g	growth rate of net annual income or dividend, where appropriate
i	interest rate
k	Project or "free and clear" IRR
N_t	net operating income for year t
q	equity IRR
R	real interest rate
R_P	portfolio return
r	rate of return or discount rate, where appropriate
r_B	cost of bond
r_D	cost of debt
r_E	cost of equity
r_F	nominal risk-free interest rate, equals $r_f + \pi$
r_f	real risk-free interest rate
r_m	market return
r_s	social opportunity cost of capital
t	time or tax rate, where appropriate, e.g. x_t, $(1 - t)$
V	gross or net value, such as share value or bond value

$E[.]$	Expectation operator; $E[x]$ is the mean of x.

β	a firm's beta if used in CAPM
μ	mean value
π	rate of inflation
σ	standard deviation or volatility (in real options)

$	Singapore dollars; all other currencies will be indicated, e.g. US$.

Preface

This book arises from my lecture notes in the Development Finance module taught to students studying for the Master of Science (Project Management) degree at the National University of Singapore (NUS). These students came from many disciplines including architecture, building, business, construction, engineering, and information technology.

Currently, there is no similar book on development finance that treats the subject from a project manager's perspective. I have planned to write this book in response to many requests from my students for a suitable textbook, and one that is concise. They find that many books in the market are repetitive. Students also like to read books that explain abstract concepts clearly.

My procrastination ended when Tony Moore, Senior Editor at Taylor & Francis Group, suggested that I should find time to start writing a book and stop preaching.

The features of the book include

- an early emphasis on discounting to provide a firm grasp of the mechanics of project financing;
- establishing the link between corporate strategy and projects;
- providing a firm foundation on full recourse corporate finance;
- a concise write-up on the project development cycle;
- benefit-cost analysis to address environmental and third-party externalities as well as market distortions in social projects;
- in-depth analysis of limited recourse project finance;
- real option pricing and other derivatives;
- a framework for risk management; and
- case studies on four different types of projects, namely, power projects, airport projects, commercial real estate developments, and chemical storage projects.

The case studies are used to illustrate aspects of large-scale infrastructure projects and focuses on how the nexus of contracts, agreements, bonds, guarantees, insurance, and other risk management strategies may not work effectively in different sectors.

The book is intended for graduate and senior undergraduate students, researchers, and parties involved in project finance including company directors, project managers, lenders, lawyers, architects, contractors, engineers, quantity surveyors, regulators, suppliers, insurers, and investors.

Friends and colleagues at the *Project Management Institute* (PMI), *International Project Management Education Union* (IPMEU), and *Center for Project Management and Construction Law* at NUS provided useful formal and informal feedback. In particular, I would like to thank the following persons and reviewers:

Navindran Davendran	KBR
Gerard De Valance	University of Technology Sydney

Jacques Desbiens	University of Quebec
Hoon Eng Eoon	Taisei Corporation
Jiang Hongbin	Asia Development Bank
James Joiner	University of Texas at Dallas
Michael Khoo	Keppel Energy
Mohan Kumaraswamy	University of Hong Kong
Terry Quanborough	TQ Projects, Sydney
Paviz Rad	Project Management Excellence, NJ
Himal Suranga	University of Moratuwa
Jim Szot	University of Texas at Dallas
Jason Teou	CapitaLand
Sam Wamuziri	Napier University

Finally, I would like to thank Katy Low for her patience and excellent editorial assistance, Sunita Jayachandran for efficient management of the production process, and Hon To for the apt image on the front cover. As usual, the errors are mine.

Willie Tan

1

Introduction

1.1 Purpose of this book

The purpose of this book is to provide a guide on the principles of project and infrastructure finance to students and practitioners.

By "principles," we mean a set of rules or claims about the nature of something. Principles tend to be general or universal, and go beyond specific ways of financing individual projects. They are developed by abstracting from "reality" (or "facts") to isolate only the essential elements for analysis. The non-essential elements are ignored. For any set of essential elements, further approximations (such as absence of friction in physics or a featureless plain in models of city form) are then made to develop general models that can subsequently be applied to specific cases.

Sometimes, the model assumptions are not approximations of reality but are quite unreal. Simple examples include the assumption of a closed economy in Keynesian economics and the assumption in neoclassical growth theory that an economy produces only one homogeneous good. In a globalized world, the assumption of a closed capitalist economy is clearly unrealistic. As for growth theory, even Robinson Crusoe produces more than one good.

However, the intent is not so much to reflect or construct "reality" but to invoke obviously unreal assumptions as a pedagogic device to start with a simple model and progressively make it more and more realistic (or "concrete") by relaxing these assumptions. Once we understand how a closed economy operates, the next step is to consider what happens in a more complicated open economy with external trade and investment.

These principles should be simple to understand and not be unnecessarily cluttered with details. For instance, one does not begin by teaching students how to solve

$$1.23x^2 + 2.234x + 23.54 = 0$$

even though it may be encountered in practice. Instead, one starts with a simpler structure such as

$$x^2 + 2x + 1 = 0$$

and explore possible solutions. The first equation is no more "practical" than the second one. From a pedagogic viewpoint, the second equation is more "practical" in teaching students how to solve quadratic equations. As is well known, there is nothing as practical as a good theory to guide us.

1.2 What is project finance?

Project finance is a form of financing a capital-intensive project (such as an infrastructure project) on non-recourse or limited recourse basis through a special project vehicle (SPV).

The recourse for lenders is primarily the revenues generated by the project for loan repayment with project assets as collateral. Unlike corporate finance, lenders do not have recourse to the assets of the sponsors (e.g. the parent corporation) should the project fail.

This special non-recourse or limited recourse feature makes project finance attractive to sponsors because financing is off-balance sheet. It does not jeopardize the parent company's ability to borrow funds for other purposes or investors' assessment of its liabilities in the balance sheet.

However, with the trend towards better corporate governance and greater transparency, corporations are generally required to report material off-balance sheet items at least in the footnotes of their annual or quarterly reports. The disclosure covers debt and guarantees as well as purchase, lease, and contingent obligations.

For lenders, pure non-recourse lending is risky and they usually require some limited form of contingent financial support from sponsors over and above their equity share as well as other forms of credit enhancements and third-party guarantees. For instance, if the borrower is a local public agency, the government may be called to guarantee repayment or provide limited contingency support if the project is delayed and requires additional funds.

Since projects are capital-intensive and interest on debt is deductible in the computation of corporate tax, project financing is highly levered to about 60–85 per cent of project cost. In many cases, equity from a sponsor or a few sponsors is insufficient because of the large investment and the desire not to make the SPV a subsidiary. Hence, the equity must be supplemented by funds from other (possibly passive) equity investors.

On the debt side, lending is often syndicated under a lead bank or arranger to pool the funds and spread the risk among a few lenders. In more lucrative projects, it may be possible to issue local, regional or global bonds (particularly when the project is near completion, thereby removing a substantial part of the construction risk) to attract funds from other investors. Once a project is completed, a permanent lender (such as a mutual fund, real estate investment trust, or insurance company) takes over the loan from the syndicate of construction lenders. Usually, securing a permanent loan first makes it easier for the borrower to obtain a construction loan. There are, of course, many other variations in project financing.

It can be seen that project finance is complicated to arrange, and this raises the issue of the cost of arranging such a loan. However, without this form of financing to pool resources and share the risk, many projects may never have gotten off the ground.

1.3 History of project finance

The origin of project finance is obscure. In Antiquity, the construction of large-scale infrastructure was largely financed by the State through taxation or looting of the assets of enemies. Constrained by inadequate finance and absence of long-term debt, many such projects were built in stages using forced and unproductive labor and took a long time to complete even if they were uninterrupted by wars and bad weather.

During the Middles Ages, merchants began raising money to finance shipping. However, because of the high risk, prospectors had only limited recourse if a ship sank. Some bridges, canals, and roads were also financed by apportioning shares to each member of the community, a procedure that could potentially lead to disputes over whether the ability to pay principle or benefit principle should apply when it came to charging fees. Money for repairs and maintenance was also raised in a similar manner, although some river and road tolls were "simply extortion" and not for improvements and were levied when "higher political authority could not prevent robber barons and local jurisdictions from levying on passersby" (Landes, 1999, p. 245). Even when tariffs were set, the toll-takers made it a point not to publish them so that they could change the levy "as opportunity offered" (p. 246). Grand churches and monasteries were also built either from church coffers or endowments. As a powerful medieval institution, the church was able to amass large tracts of land bequeathed by childless widows of warring knights killed in battles.

With the advent of capitalism, money was also raised to finance railways and real estate in places such as India, Africa, the Americas, Malaya, Australia, and New Zealand in the 19th century. For short railways, finance was arranged locally normally through share issues to friends, farmers, manufacturers, miners, and other prospectors. Information on the new railway technology, location of minerals, soil conditions, reputation of promoters, and labor problems were too sparse for these speculative projects to attract external finance. Larger projects were able to attract the interests of investment houses after colonial States were asked to guarantee minimal rates of return (Thorner, 1951). That is, if a project did not provide sufficient return on a bond, the State would pay the difference between the declared return (e.g. 8 per cent) and what the project company paid out (e.g. 5 per cent). Many of these State-backed bonds were floated in London to tap surplus British capital. Typically, investment houses bought shares or bonds to signal confidence in their advice to British investors.

As cities boomed, developers began to set up separate special purpose vehicles to raise funds and protect their liabilities should a project fail. To limit their equity, end-user finance was also used in the form of progress payments from buyers of real estate under construction. Many of these projects were speculative in the sense that they were built in anticipation of demand rather than at the request of home owners. Since such demands were cyclical and volatile, the risks and pay-offs were high.

After World War II, large-scale infrastructure was largely financed by the State through tax revenues, borrowings, or simply over-printing of money by tolerating "some" inflation. This expansionary approach was in line with the idea of State-led development using State planning and Keynesian deficit finance to raise aggregate demand during a downturn and achieve full employment. As the economy recovers, tax revenues will rise and State expenditure can be scaled back. In theory, the budget surplus during a boom will be used to cover the deficits incurred during a downturn. In practice, government spending can spin out of control as politicians seek re-election and make too many promises.

The classic case for State subsidy of large-scale infrastructure projects rests with the external benefits brought about by the improved infrastructure. In addition, in the 1950s and 1960s, few private firms were able to undertake such huge and lumpy projects without State assistance or assurances against confiscation of project assets or changes in taxes or regulation.

As we have just seen, the subsidy took the form of State guarantee of the interest on bonds. State guarantees ensured that projects were attractive enough to be financed externally but it also weakened the profit incentive for promoters and sponsors. It encouraged mismanagement, looting, over-promotion, inflated prices, and ruinous competition. Not surprising, there were spectacular failures (Grodinsky, 2000; Nairn, 2002).

In the developing countries, domestic savings were supplemented by foreign aid and funds from lending international agencies. Since

$$y = C + I + G + X - M = C + S + T,$$
$$S - I = (G - T) + (X - M). \tag{1.1}$$

Here y is real gross domestic product or national income. It is equal to the sum of private consumption expenditure (C), business fixed investment and residential investment expenditure (I), government expenditure (G), and net exports (exports less imports, or $X - M$). The national income is also the sum of consumption expenditure, savings (S), and taxes (T). Equation (1.1) shows that there are three well-known "gaps" to bridge in economic development (Chenery and Bruno, 1962):

- a "savings gap" in the private sector,
- a government budget deficit, and
- a trade deficit.

These gaps may be "plugged" by mobilizing domestic savings, balancing the State budget, and overseas borrowings or through external aid.

At the local government level, tax-exempt revenue bonds were used to finance the revitalization of many cities (particularly in the United States) through urban renewal. This method is still in use; the destruction of the American Gulf Coast by hurricanes Katrina and Rita in 2005 also led to bond issues based on utility tariffs rather than assets to rebuild the infrastructure.

In the energy sector, private firms began to use project finance to tap rapidly developing capital markets for mineral, oil, and gas exploration (e.g. North Sea oilfields). In addition to equity and traditional bank lending, international bonds were also used to raise large sums of money simultaneously worldwide. During the inflationary 1970s and early 1980s, debt financing was attractive to investors since repayments were made in cheaper dollars.

In the 1990s, the securitization of assets became popular to "unlock" asset value and provide liquidity to owners. Many large commercial developments were securitized and bought by Real Estate Investment Trusts (REITs). Dividends from these trusts were attractive to many small investors because of tax benefits. For instance, in the US, REITs are not taxed on the income distributed to shareholders if at least 90 per cent of the ordinary taxable income is distributed annually as dividends. This avoids the usual double corporate taxation on profits and dividends paid to shareholders.

The swing against "big" governments towards free markets for greater efficiency and transparency led to the development of public-private partnership (PPP) projects. As we saw earlier, large-scale infrastructure projects from the 1950s to the 1970s were largely owned by the government and funded from domestic savings, taxation, and overseas borrowings or through foreign aid. This strained the budgets of many governments, and PPP projects were conceived as a way for the State to partner the private sector in developing such projects. The private sector provides the badly needed funds and expertise.

1.4 Approaches to project finance

Generally, approaches to project finance may be categorized as follows:

- *procedural*, such as Pahwa's (1991) book on *Project Financing*. The 1136-page book is replete with policies, rules, forms, annexes, and checklists;
- *case study-centered*, for example, Lang's (1998) book on *Project finance in Asia* and Esty's (2004) *Modern project finance: a case book*;
- *finance-centered*, such as Finnerty's (1996) *Project financing* and Nevitt and Fabozzi's (2000) *Project financing*;
- *legal*, for instance, Vintner's (1998) *Project finance* and Hoffman's (2001) *The law and business of international project finance*;
- *integrated*, where attempt is made to integrate the financial, engineering, economic, environmental, and legal aspects of project finance. This is the aim of Khan and Parra's (2003) book on *Financing large projects*. Such a comprehensive approach is more a desired state than reality. There is little discussion on the financial, engineering or environmental aspects of projects; and
- *analytical*, where the concern is more academic, and the findings are published in academic journals (e.g. Shah and Thakor, 1987; Chemmanur and John, 1996).

This book uses a *managerial approach* based on my lectures in teaching project managers from industries such as information technology, engineering, construction, manufacturing, and oil and gas. It is neither a special area treatise in law or finance, or a book on procedures or a collection of academic articles. Neither does it focus exclusively on case studies nor tries to integrate all aspects of project finance.

Project managers should understand the basic principles of finance, the main features of project finance, and possess the analytical skills to apply these principles to real projects in a holistic way. In short, it places project finance in the context of project management practices.

1.5 Importance of project finance

Why should managers of large projects learn about project finance? There are many reasons. First, if money makes the world goes round, it drives large projects as well. Many large projects will never get off the ground if they cannot attract sufficient internal or external financing. Since projects need to be prioritized because of budgetary constraints, it is essential for project managers to know how to package a financially viable project for approval. In particular, it is important to know the sources and cost of project funds, taking into consideration complex tax regimes.

Second, these projects are subjected to large financial risks in the form of changing interest rates on long-term loans and volatile currency markets that affect loan repayments, input supply prices, and output prices. These risks need to be hedged.

Third, a project's cash flows need to be managed properly. These cash flows must be planned and anticipated to allow sponsors to arrange for the requisite funds. Project managers must appreciate that delays translate into large sums of money.

Finally, lenders impose loan covenants as a condition for lending. Since these covenants constrain project performance, it is necessary for project managers to understand such covenants. Generally, these covenants cover issues such as project performance as a condition for disbursement of funds, separate accounts to monitor the movement of funds, periodic reporting, and prior approval before raising additional funds.

1.6 Organization of this book

The book is organized as follows.

Since time is money, the basics of the time value of money of discounting should be mastered at the earliest opportunity (Chapter 2). Discounting is used throughout this book.

Projects are undertaken by public and private organizations as part of corporate and business strategies, and project managers need to understand the strategic and business side of project finance (Chapter 3). In addition, a good grasp of corporate finance (Chapters 4 and 5) is essential to understand the options and constraints in financing projects.

Chapter 6 provides a brief discussion on the project cycle from inception to project close-out. This is followed by additional complications in dealing with social projects (Chapter 7). Unlike private projects where profitability is based solely on revenues less costs, the feasibility of social projects involves consumers' surplus, externalities, and other distortions that are not captured in the profit calculus of private projects.

With these preliminaries, the structure of project finance in terms of its stakeholders is considered in Chapter 8. Since the nexus of contracts deals primarily with risk, an introductory risk management framework is provided in Chapter 9. This is followed by considering the relation between risk, insurance, and bonds (Chapter 10) and financial risks (Chapters 11 and 12). Specific provisions in contracts and agreements against the various types of risks are discussed in Chapter 13.

The last four chapters contain case studies on various types of projects, namely, electric power, airport development, real estate development, and chemical storage projects. These projects are based on different aspects of actual cases using publicly available information. They are discussed more broadly as "projects" rather than a single case study as a pedagogic device to draw out useful lessons. The aim is not to repeat the issues discussed in the earlier chapters but to illustrate how the nexus of contracts, agreements, bonds, and other risk management devices may break down or operate in unexpected ways, or to discuss novel features of a project or different types of projects.

Questions

1   What are the key features of project finance?

2   How does project finance differ from corporate finance?

3   Why is the study of project finance important?

4   What is the rationale for State subsidy or guarantee of infrastructure projects? What are the risks in providing such subsidies or guarantees?

5   Explain why the State in less developed countries (LDCs) may be short of funds to invest in infrastructure. How can this problem be tackled?

6   What accounts for the recent popularity of public-private partnerships in structuring projects?

2

Time Value of Money

2.1 Future value of present sum

A solid foundation in the time value of money is a pre-requisite in understanding project finance. Large infrastructure projects are implemented over long periods of time lasting from a few years to even decades. Normally, large projects are executed in phases.

Over time, the value of money differs because of

- the risk of default,
- inflation, and
- the time the money could be put to productive use (i.e. the opportunity cost of money).

Hence, borrowing comes at a cost, and this cost or "price" of borrowing money is the rate of interest. A dollar today is worth more than a dollar next year depending on the prevailing rate of interest.

What is called "the" rate of interest is an abstraction. In practice, there are many rates of interest such as deposit rates, credit card interest rates, mortgage interest rates, and prime lending rates on loans to corporations. These rates also differ across banks. Hence, the use of "the" interest rate is merely a way to simplify the large variety of interest rates. Sometimes, it is useful to think of it as the general level of interest rate in the economy. Even here, "the general level of interest rate," like the consumer price index, is a theoretical construction. Indexes can be good or bad depending on the theory used to construct them. Fortunately, we do not need to know how to construct the general level of interest rate to understand the time value of money.

The mechanics of discounting future values to present values is relatively simple. The trick to mastering it is to understand the basic principles rather than remember formulas, look up tables, solve uninteresting puzzles, and crunch numbers. For this reason, financial tables or financial calculators will not be used here. These "tools," popular with Finance 101 books, tend to impede rather than enhance understanding. A novice is easily overwhelmed by the large number of formulas when, in fact, a few basic principles is all that is required.

If a sum P is invested at interest rate i for a year, its *future value* one year later is

$$P_1 = P(1 + i).$$

If the sum is reinvested for another year, its value at the *end* of the second year is

$$P_2 = P_1(1 + i) = P(1 + i)^2.$$

In general, the future value at the end of year t is

$$P_t = P(1 + i)^t. \tag{2.1}$$

Example

What is the future value of $200 at the end of three years if it is invested at 5 per cent interest?

Here $P_3 = P(1 + i)^t = 200(1 + 0.05)^3 = \$231.53$.

2.2 Present value of future sum

From Equation (2.1), the *present value* of a future sum is given by

$$P = \frac{P_t}{(1+i)^t}. \tag{2.2}$$

Example

If you are to receive $100 at the end of 3 years from now, what is its present value, assuming a 6 per cent interest rate?

Here $P = \dfrac{P_t}{(1+i)^t} = \dfrac{100}{(1+0.06)^3} = \$83.96$.

2.3 Present value of income stream

An asset such as a bond, house or facility earns a dividend, net rent, or net operating income each year. Like net rent, net operating income is gross income less operating expenses. Its present value V is therefore the sum of all future incomes discounted to present value at the appropriate discount rate. That is, for a facility,

$$V = \frac{N_1}{(1+i)} + \frac{N_2}{(1+i)^2} + \cdots + \frac{N_n}{(1+i)^n} \qquad (2.3)$$

where N_t is the net operating income at the end of year t, $t = 1,\ldots, n$. For a financial asset such as a bond, N is the annual dividend.

This equation is the workhorse of the time value of money, and variations in the formula relate to differing assumptions regarding the Ns, n, and discounting period (which so far has been assumed to be yearly). Some of these variations are discussed below.

Note the inverse relation between the value of an asset and interest rate. If interest rate rises, the value of an asset falls if its income stream is constant. For instance, if interest rate rises, the prices of bonds fall.

Example

If a house can be net rented for $10,000 a year and has a remaining lease of 4 years, what is its present value, assuming a discount rate of 5 per cent?

Using Equation (2.3),

$$\begin{aligned} V &= \frac{N_1}{(1+i)} + \frac{N_2}{(1+i)^2} + \cdots + \frac{N_n}{(1+i)^n} \\ &= \frac{10,000}{(1.05)} + \frac{10,000}{(1.05)^2} + \cdots + \frac{10,000}{(1.05)^4} = \$35,460. \end{aligned}$$

Example

What is its present value if the above house is a freehold asset that earns a net rent of $10,000 in perpetuity?

From Equation (2.3),

$$\begin{aligned} V &= \frac{N_1}{(1+i)} + \frac{N_2}{(1+i)^2} + \cdots \\ &= N\left[\frac{1}{(1+i)} + \frac{1}{(1+i)^2} + \cdots\right] \\ &= \frac{N}{i} = \frac{10,000}{0.05} = \$200,000. \end{aligned} \qquad (2.4)$$

The second line is obtained by noting that

$$\frac{1}{(1+i)} + \frac{1}{(1+i)^2} + \cdots = \frac{1}{i}.$$ (2.5)

2.4 Gordon's formula

Gordon's formula is also a variation of Equation (2.3) where the freehold asset is assumed to earn a net annual income (N) that grows at g per cent per year. Then Equation (2.3) becomes

$$
\begin{aligned}
V &= \frac{N(1+g)}{(1+i)} + \frac{N(1+g)^2}{(1+i)^2} + \cdots \\
&= N[a + a^2 + \cdots + a^t + \cdots] \qquad \text{where } a = (1+g)/(1+i) \\
&= \frac{N(1+g)}{(i-g)} = \frac{N_1}{i-g}.
\end{aligned}
$$ (2.6)

Proof

To evaluate the sum in [.] in the second line, let

$$S_n = a + a^2 + \cdots + a^n$$

so that

$$aS_n = a^2 + \cdots + a^{n+1}.$$

Subtraction gives

$$S_n - aS_n = a - a^{n+1}.$$

That is,

$$S_n = \frac{a - a^{n+1}}{1-a}.$$

Since

$$a = \frac{1+g}{1+i},$$

a^{n+1} tends towards zero as n tends towards infinity, provided $g < i$. That is, the rate of income growth must be less than the discount rate. Otherwise, the sum S_n is "explosive" and the value of the asset is no longer finite. Dropping the subscript n from S_n to indicate the sum of infinite terms, we have

$$S = \frac{a}{1-a} = \frac{1+g}{i-g}$$

since $a = \dfrac{1+g}{1+i}$. Thus,

$$V = \frac{N(1+g)}{i-g}.$$

Letting $N_1 = N(1 + g)$ gives Gordon's formula. This completes the proof.

Example

A share pays a current dividend of \$2 (= N_1). If the discount rate is 5 per cent and dividends are expected to grow at 2 per cent a year (i.e. $g = 0.02$) till perpetuity, what is the fair value of the share?

Here $V = \dfrac{N_1}{i-g} = \dfrac{2}{0.05-0.02} = \$66.67.$

Example

If a firm's share is currently trading at \$4 per share and the dividend of \$0.25 is expected to grow at 1% annually, what is the rate of return required by equity investors?

Rearranging Gordon's formula gives

$$i = g + \frac{N_1}{V}$$

$$= 0.01 + 0.25/4 = 0.0725 \text{ or } 7.25\%. \qquad (2.7)$$

The weaknesses of Gordon's formula include using only the first year dividend and historic dividends to estimate g as inputs into the formula to compute the rate of return to equity.

2.5 Real and nominal rates of interest

Recall that if a sum P is invested at interest rate i for a year, its future value one year later is

$$P_1 = P(1 + i). \tag{2.8}$$

If the real interest rate is R, then

$$P_1 = P(1 + R)(1 + \pi) \tag{2.9}$$

where π is the expected rate of inflation. Equating Equations (2.8) and (2.9),

$$\begin{aligned}
1 + i &= (1 + R)(1 + \pi) \qquad \text{(Fisher relation)} \\
&= 1 + R + \pi + R\pi \\
&\approx 1 + R + \pi.
\end{aligned} \tag{2.10}$$

Hence,

$$R = \frac{1+i}{1+\pi} - 1 \tag{2.11}$$

if the first line of Equation (2.10) is used, and

$$R \approx i - \pi \tag{2.12}$$

if the last line of Equation (2.10) is used. The approximation is good enough in most cases.

Example

What is the real rate of interest if the nominal rate is 7 per cent and the expected rate of inflation is 3 per cent?

The exact solution is

$$R = \frac{1+i}{1+\pi} - 1 = \frac{1.07}{1.03} - 1 = 3.9\%.$$

The approximate solution is

$$R = 7\% - 3\% = 4\%.$$

The approximation is close enough for most practical purposes.

Example

What will happen to nominal interest rates if inflation is expected to rise substantially?

From Equation (2.10),

$$i = R + \pi.$$

If the real interest rate is stable, nominal rates will rise in tandem with the expected rate of inflation.

2.6 Components of interest rates

From Equation (2.12),

$$i = R + \pi$$
$$= r_f + \lambda + \pi$$
$$= r_F + \lambda \qquad\qquad (2.13)$$

where the real interest rate (R) is broken up into a real risk-free rate (r_f) to compensate investors for parting with liquidity or postponing consumption, and a risk premium (λ) to compensate investors for investment-specific risks including the risk of default. The last line is obtained by setting

$$r_F = r_f + \pi,$$

that is, nominal risk-free rate equals real risk-free rate plus the expected rate of inflation.

Equation (2.13) implies that the interest rate consists of a risk-free rate plus a risk premium. The risk premium differs across asset classes (Figure 2.1) and projects. Among asset classes, bank deposits and corporate bonds are generally safer investments than property and stocks. Hence, a higher risk premium in property and stock investments is required to compensate investors for the risk. The expected returns on property and stocks are therefore higher than that of deposits and bonds but so are the risks. These risks may be approximated by the standard deviation of returns.

It will be shown in Chapter 5 that, for an individual stock, the expected return is given by

$$E[r] = r_F + \lambda = r_F + \beta[E(r_m) - r_F] \qquad\qquad (2.14)$$

where β is a firm-specific constant (called a firm's beta), and r_m is the market return. Equation (2.14) provides an explicit expression for the risk premium λ. The important point here is that the expected return on a stock is linearly related to the excess market return (i.e. $E(r_m) - r_F$). Equation (2.14) has a special name; it is called the *security market line*.

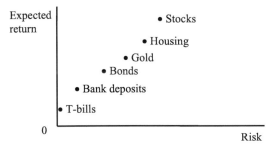

Figure 2.1 Risk and return for various classes of assets.

2.7 Determinants of interest rates

In the long run, the economy-wide interest rate (i) is determined by the demand and supply of funds by firms, households, and governments (Figure 2.2).

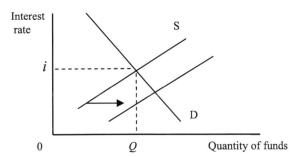

Figure 2.2 Determination of interest rate.

The supply of funds depends on factors such as the interest rate, household and corporate income (Y), tax rate (t), government debt or surplus (G), and monetary policy (M). The supply of foreign funds is largely a function of interest rate and can be left out of the supply *function*

$$S = S(i, Y, t, G, M).   \tag{2.15}$$

Since it is difficult to plot this supply function, we hold all other variables except interest rate constant to obtain the supply *curve*

$$S = S(i).   \tag{2.16}$$

The supply curve is a function on interest rate only, and is upward-sloping because the higher the interest rate, the greater is the incentive to save. If income or the tax rate changes (i.e. other variables are not constant), the supply curve will shift left or right. For instance, higher incomes tend to encourage more savings, resulting in a shift of the supply curve to the right as shown in Figure 2.2.

The demand for funds depends on the interest rate, government surplus/deficit G, and general economic outlook. The latter may be proxied by national income (y). If the business outlook is good, firms increase their investment and demand for funds. Similarly, households borrow more to finance their purchases of houses, cars, and so on. Hence, one can write the demand function

$$D = D(i, G, y).   \tag{2.17}$$

Holding G and y constant gives the demand curve

$$D = D(i).   \tag{2.18}$$

The demand curve is shown in Figure 2.2.

With lower interest rates, more projects become viable, and governments and firms tend to borrow more funds to carry out their private and social projects. Consumers will also borrow more to fund their purchases. Hence, the demand curve is downward-sloping.

The equilibrium interest rate is determined by the demand and supply of funds. Equating Equations (2.15) and (2.17) gives

$$i = i(G, y, Y, t, M).   \tag{2.19}$$

Thus, interest rates are determined by government expenditure (G), national income (y), household and corporate income (Y), the tax rate (t), and monetary policy (M). The latter affects interest rates through the banking system. If money supply is defined as cash and bank deposits, then the Central Bank can influence the amount of funds banks can lend by either varying their cash holdings or deposits. Banks hold only a small amount of cash to meet short-term withdrawals (called fractional banking), and the rest of the deposits are lent out. By varying the amount of money banks must hold as cash (called the Legal Reserve Ratio or LRR), the Central Bank limits the amount of loanable funds.

In practice, the LRR is seldom varied. A more effective way of influencing bank lending is to influence deposits rather than the small amount of cash banks hold. One method of doing this is through open market operations (the buying and selling of short-term treasury bills and long-term government bonds). If the Central Bank buys back its bonds from the public, it issues checks and these checks enter the banking system when holders deposit them with their banks. In turn, banks have more funds to lend to other customers.

The other way the Central Bank can influence money supply is through the discount rate (not to be confused with the discount rate in computing present value), the rate it charges commercial banks. This is because banks borrow and lend to each other at the interbank rate such as the London Interbank Offered Rate (LIBOR), Singapore Interbank Offered Rate (SIBOR), or federal funds rate in the US. If the discount rate is raised, bank borrowings become more expensive, leaving banks with less loanable funds and the higher interest rate is passed to customers. Since banks can borrow money elsewhere, changing the discount rate to implement monetary policy is generally less effective than altering the federal funds rate through open market operations.

In the short run, Keynes (1936) argued that interest rates are less determined by the demand and supply of loanable funds outlined above than by liquidity preference. According to Keynes, the interest rate is not primarily determined by savings or consumption decisions but by the portfolio decision whether to hold money as an asset. Hence the interest rate is seen as the reward for parting with liquidity and its level is determined in the money market. Keynes argued that money is demanded for transaction, precautionary, and speculative motives, and he stressed the speculative motive. If interest rates are low, bond prices are high (see Equation (2.3) on the inverse relation between asset values and interest rates) and investors expect bond prices to fall. Hence, they sell bonds, that is, prefer liquidity and the demand for money is relatively insensitive to interest rates at this low level. This relatively flat portion of the demand for money curve (M_D) is called the "liquidity trap" (Figure 2.3). In the short run, the supply of money (M_S) is assumed to be fixed by the Central Bank and is drawn vertical as shown. Note that Keynes used the term "bonds" in a generic sense to refer to financial assets.

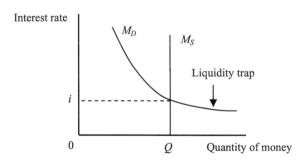

Figure 2.3 Keynesian theory of interest.

Apart from rejecting the idea that interest rate is the reward for waiting, Keynes further argued that the loanable funds theory is indeterminate. This is because savings depends on income and the latter depends on investment that, in turn, depends on the interest rate.

In summary, there are two views on how interest rates are determined. The neoclassical view is based on the demand and supply of loanable funds and because these are based on savings and consumption decisions, they tend to operate in the long run. The Keynesian view is that, in the short run, the interest rate is not primarily determined by the decisions to consume or save but by portfolio considerations on whether to hold money as an asset given its rate of return (the interest rate). Keynes is probably right that savings and investment are not sensitive to interest rates in the short run. However, in the long run, the level of interest rate will depend on the level of savings.

In practice, a Central Bank either targets interest rates or money supply to control inflation and achieve other objectives of monetary policy (e.g. foster economic growth). It has been found that targeting monetary aggregates are more difficult to implement so the preference in the US is to target the federal funds rate (i) through open market operations using Taylor's (1993) rule:

$$i = n + \pi^* + \gamma(\pi - n) + \phi y \qquad (2.20)$$
$$= 2 + \pi^* + 0.5(\pi - 2) + 0.5y \quad \text{(based on empirical studies)}$$

where

n = equilibrium or "natural" real interest rate (2 per cent say);
π^* = target inflation rate (2 per cent say);
$y = 100(Y - Y^*)/Y^*$ = measure of gap between real potential (trend) GDP (Y^*)
and real GDP (Y); and

γ, ϕ = parameters.

The rule is not strictly applied; rather, it is used to guide monetary policy. For instance, if the economy is near full employment level or inflation is high, the rule recommends raising interest rates. Conversely, interest rates should be lowered if the economy is in a recession or if inflation is low. For example, if the rate of inflation is 3 per cent and the economy is operating at full employment (i.e. $y = 0$), then

$$i = 2 + 2 + 0.5(3 - 2) = 4.5\%.$$

2.8 Term structure of interest rates

The above exposition on the determination of interest rates is static since it determines the rate of interest only at a point in time. In practice, the differing maturities on financial instruments, all else equal, result in different interest rates.

Thus, shorter-term bonds yield lower interest rates than longer-term bonds, and this variation of interest rate with maturity, when plotted, is called the yield curve (Figure 2.4).

The higher interest rate on longer-term bonds compensates holders for inflation risk. That is, the holder of a long-term bond is "locked in" and faces a greater risk of higher future inflation than the holder of a short-term bond. Tax differences can also result in differing bond yields with maturities by changing the net yield of bonds.

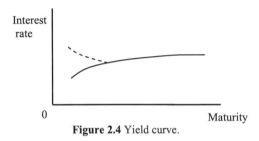

Figure 2.4 Yield curve.

It is also possible for short-term rates to be temporarily higher than long-term rates during periods of credit tightening when funds are in short supply. Interest rates are then lower on longer-term bonds because investors expect monetary policy to ease. Hence, the yield curve is "inverted" as shown by the dotted line in Figure 2.4.

2.9 Deficit financing and interest rates

We have seen how a government surplus or deficit affects the supply and demand for loanable funds and hence the interest rate.

There are many ways of financing a budget deficit. If the State raises taxes, then households and firms will have less money to save in the short run. Consequently, the supply curve for loanable funds shifts to the left and, if the demand curve for funds remains unchanged, interest rates will rise.

Alternatively, the government may borrow from the capital market by selling bonds to raise the money. The public will have less money after buying the bonds and the supply curve for loanable funds shifts to the left. Consequently, all else equal, interest rates rise. Government borrowing is then said to "crowd out" less profitable private investment or projects by raising interest rates. Projects with rates of return lower than the prevailing interest rate will then no longer be viable. However, the rise in interest rates attracts funds from abroad and therefore partially offsets the domestic crowding out effect. Consequently, empirical evidence on the crowding out hypothesis is mixed.

Finally, if the government prints money to pay off its debt, the cash will be deposited into the banking system. In the short run, there is more money in the banking system and interest rates will tend to fall. However, in the long run, there will be too much money chasing after too few goods if the economy is at full employment level, and this method of financing a deficit is inflationary. This line of reasoning is based on the classical Quantity Theory of Money where

$$MV = PQ.$$

Here, M is money supply, V is the number of times money is used (velocity of circulation), P is the general price level, and Q is real output. In the long run, the economy is assumed to be at full employment level so that Q is fixed. Similarly, V is assumed to be fairly constant. To see this, we can write

$$V = PQ/M,$$

that is, velocity is a fairly constant ratio of the value of output to money supply because of the way money is used in the economy as a medium of exchange, a unit of account or a store of value. A simple analogy is to say people tend to have a fairly constant amount of money in their wallets for their daily needs. Since V and Q are fairly constant,

$$M = (Q/V)P = kP$$

where $k = Q/V$ is relatively fixed. Consequently,

$$\Delta M = k\Delta P,$$

that is, a change in money supply leads directly to a change in the general price level in the long run. In essence, this is the monetarist theory of inflation.

2.10 Credit rationing

We have seen how the equilibrium interest rate is determined by the demand and supply of loanable funds. However, not all projects are financed at the equilibrium interest rate.

In Figure 2.5, the interest rate is "repressed" at disequilibrium rate p to encourage investment in projects by lowering the cost of borrowing. Under such a regime, commercial banks were often nationalized and politically directed to lend to "priority projects" at relatively lower repressed rates. In addition, State-owned rural, construction, and industrial banks also established special credit schemes to channel funds to selected projects. Amsden (1992) has argued that such "developmental States" in countries such as South Korea and Japan deliberately "got prices wrong" (note that interest rate is the price of using money) to promote industrialization by shifting funds to priority sectors.

However, at the repressed rate, demand for funds exceeds supply ($Q_D > Q_S$). At this rate, savings tend to dry up while the demand for funds tends to grow as projects with lower rates of return become viable. Hence, some form of (disequilibrium) credit rationing is required to lend out the limited funds. The unsatisfied demand is diverted to the unofficial "curb market" where lending takes place at much higher interest rates (c).

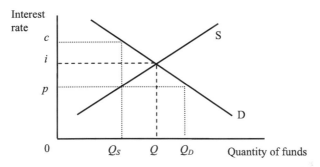

Figure 2.5 Financial repression.

The highly differentiated interest rates resulted in allegations about political favoritism and misallocation of resources through over-investment in projects with low returns as well as over-investment in capital because of lower interest rates, thereby raising the capital intensity of production. Consequently, many financial markets were liberalized in the 1980s and 1990s to "get prices right" as well as "get policies right."

In practice, credit rationing occurs all the time, even if the interest rate is at equilibrium level. Risky borrowers may not be able to borrow at interest rate i, or any other interest rate. Raising the interest rate may not be the solution since risky borrowers may be tempted to try their luck and walk away from the project if it fails. Credit rationing also occurs because of the lumpiness of projects. For example, if $10 bn is required, lenders may only lend out $8 bn not only because of project risk but also because the cost of funds (supply curve) rises beyond the equilibrium interest rate level beyond Q (Figure 2.6). Hence, $Q_D - Q$ is rationed out.

Fixing interest rates below equilibrium levels is just one form of financial repression. The other forms include

- capital controls on residents holding foreign assets or domestic firms borrowing abroad;
- restrictions on entry into the financial sector, leading to limited competition; and
- high reserve requirements and liquidity ratios imposed on banks.

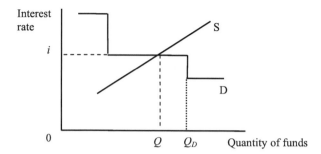

Figure 2.6 Equilibrium credit rationing.

We have seen that below-market interest rates and lending to "priority sectors" tend to misallocate resources. Controls on capital mobility will also impede market adjustment towards equilibrium, preventing firms from borrowing abroad at lower interest rates while simultaneously creating additional domestic demand for investible funds. In general, limited competition in the financial sector tends to lead to inefficiencies and higher prices. Finally, high reserve requirements and liquidity ratios slow the expansion of money supply.

Financial repression is a matter of degree. It was mild in the dynamic East Asian economies and therefore did not hamper growth prior to liberalization of financial markets in the 1990s (World Bank, 1993). Further, savings rates in these countries were relatively high despite the low deposit rates and funds were productively invested by the private and public sectors. Since priority sectors had to perform satisfactorily to continue to receive funding, there were fewer cases of misallocation in these countries.

2.11 Continuous time discounting

Recall from Equation (2.2) that if the annual rate of interest is i, the present value of a future sum at time t is given by

$$P = \frac{P_t}{(1+i)^t}.$$

For analytical purposes, it may be more convenient to compound at periods shorter than a year. For semi-annual compounding, the annual interest rate is divided by two and the number of periods is doubled so that

$$P = \frac{P_t}{[1+(i/2)]^{2t}}.$$

If compounded monthly,

$$P = \frac{P_t}{[1+(i/12)]^{12t}}.$$

If compounded daily,

$$P = \frac{P_t}{[1+(i/365)]^{365t}}.$$

More generally, if we compound n times a year, the present value is

$$P = \frac{P_t}{[1+(i/n)]^{nt}}.$$

If we compound at even shorter periods (e.g. hourly), then n gets larger and, in the limit as n tends towards infinity,

$$P = \lim \frac{P_t}{[1+(i/n)]^{nt}} = P_t e^{-it} \tag{2.21}$$

where $e = 2.71828...$ is Euler's number or base of natural logarithm. The result uses the relation

$$\lim[1+(1/x)]^x = e.$$

This can be shown numerically by using different values of x:

x	2	10	100	10,000
$[1+(1/x)]^x$	2.3	2.6	2.71	2.718

As x gets larger, the value of $[1+(1/x)]^x$ approaches e. Putting $x = n/i$ in Equation (2.21) gives the result.

Example

If you are to receive \$100 at the end of 3 years from now, what is its present value, assuming $i = 6$ per cent?

If discrete annual compounding is used,

$$P = \frac{P_t}{(1+i)^t} = \frac{100}{(1+0.06)^3} = \$83.96.$$

If continuous compounding is used,

$$P = P_f e^{-it} = 100e^{-0.06(3)} = \$83.53.$$

The answers are quite close.

Example

Derive the continuous time version of Gordon's formula.

From Equation (2.6), the discrete time version of Gordon's formula is

$$V = \frac{N(1+g)}{(1+i)} + \frac{N(1+g)^2}{(1+i)^2} + \cdots = \frac{N_1}{i-g}.$$

In continuous time,

$$V = \int_0^\infty Ne^{gt}e^{-it}\,dt = N\int_0^\infty e^{(g-i)t}\,dt$$

$$= \frac{N}{g-i}\left[e^{(g-i)t}\right]_0^\infty = \frac{N}{i-g}.$$

The answers differ slightly. In continuous compounding, the numerator is N, not N_1. This is because the value of N at the end (rather than the beginning) of year 1 is used in discrete compounding.

Questions

1   Compute the present value of $1,000 receivable in 4 years' time if the discount rate is 6 per cent. [$792]

2   If your laptop earns a net rent of $1,000 for 3 years and has a salvage value of $200 at the end of 3 years, what is its present value if current interest rate is 5 per cent? [$2,896]

3   Prove Equation (2.5), that is, $1/i = [1/(1 + i) + 1/(1 + i)^2 + \cdots]$.

4   Explain why the assumption $g < i$ is required in deriving Gordon's formula in both discrete and continuous compounding.

5   A friend offers to sell you a freehold property that currently earns a net rent of $10,000 a year and rents are expected to grow at 2 per cent annually. What is its fair value if the current interest rate is 5 per cent? [$333,333]

6   Banks in a less developed country are willing to give business loans in local currency at 14 per cent interest. If the annual inflation rate is about 9 per cent, what are the exact and approximate real rates of interest? [4.59%; 5%]

7   With the rise in oil prices, the US inflation rate is likely to rise to 4.5 per cent over the next quarter. If the US economy is operating at 98 per cent capacity, what is the target interest rate? [4.25%]

8   If a government prints money instead of raising taxes or borrowing to fund a major project, what are the economic consequences?

3

Organizations and Projects

3.1 Functions of management

A manager's role is to plan, organize, lead, and control people, activities, and processes (Figure 3.1). This view of management is based on a systems approach where a manager performs these roles to achieve certain organizational (system) goals and objectives.

The system is open, in the sense that an organization needs to take into account the opportunities and threats in its external environment if it is to survive. In other words, an organization must adapt to environment circumstances (in the same way as animals adapt to their environments) or create and exploit new opportunities. Consequently, contrary to classical management theory, there is no "one best way" of managing an organization, and different environments will tend to generate different types of organizations and dissimilar styles of management. A style of management that is effective in a particular environment may not work in another situation.

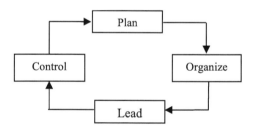

Figure 3.1 Functions of management.

To plan is to determine the organizational goals (i.e. vision and mission), tasks, and means (including staffing) to achieve these goals. Importantly, these goals may conflict with individual or group needs and objectives, and it is vital to obtain some level of consensus if the plan is to work. The manager then organizes activities by designing an appropriate organizational structure to indicate who decides, who performs, and the reporting structure.

The next managerial task is to lead (or direct) by example and by motivating people to perform work through a proper set of incentives and desired corporate culture or, more simply, the "preferred ways of doing things" within the organization.

Finally, the manager controls performance through continuous monitoring and periodic reviews of performance. A manager is proactive and takes corrective or preventive actions.

3.2 Management competencies

A manager needs to be competent in executing all the four basic functions discussed above. To be able to plan well, managers need strategic vision and decision-making competency. They must understand both their internal organizations and the rapidly changing external environment.

To be able to organize effectively, a manager needs to understand organizational design, why organizations change, how organizations "learn," and how to manage human resources.

As a leader, a manager manages her own development, is disciplined, has integrity, takes responsibility (rather than pass the buck), motivates people, communicates clearly, and understands cultural differences.

Finally, a manager needs to know how to control behaviors, activities, and processes towards desired outcomes. Preventive and corrective actions will need to be taken. The organization is not a machine, organism, or system. It is built on *people* who have different needs, values, and goals. These people may not be too keen on trite business ideas.

Details of these functions may be found in textbooks on general management (e.g. Hellriegel et al., 2005). In this chapter, the focus is on the planning function and its relation to project finance. The reason for focusing on planning is simple; project financing is largely a front-end activity. If financing cannot be structured, there is no project to manage.

3.3 Corporate strategy

A private or public organization requires a strategy or "game plan" on how it intends to succeed in its environment and achieve its goals. Nowadays, there seems to be more similarities than differences between private and public organizations. Both types of organizations have corporate strategies, use similar tools to strive for efficiency and quality, but generally differ on the profit motive.

We begin our discussion primarily with private corporate strategies, bearing in mind that many of these concepts are applicable to public organizations and have been substantially borrowed with little modifications. We shall then consider some characteristics of public organizations towards the end of this chapter. There are many books on corporate strategies, and what follows is a summary of the main ideas. Sadly, the development of corporate strategy in organizations can be rather mechanical, routine, "paper work," or a sheer waste of time. If this describes the state of affairs in your organization, then corporate strategy has not been properly articulated, planned, and executed. Employees then become cynical of (poor) management and meaningless or tiresome corporate missions and visions. As discussed below, it is important to be absolutely clear about the role of corporate strategy, its articulation, and execution. The common sources of confusion in discussions of corporate strategy include the failure to distinguish short-term and long-term strategies, demand-side and supply-side strategies, and strategies applied at different levels.

In the short term, an organization focuses on the demand side, that is, how it intends to competitively *position* itself in the market in terms of its products, geographical spread, and market segment (Porter, 1980). For instance, a budget airline may provide air travel (product) within Asia (geographical spread) for budget travelers (segment). Where possible, a firm should avoid head-on ruinous or cut-throat price competition (often called a "race to the bottom") unless it has a major cost advantage or superior quality. Instead, the firm should identify appropriate products or services, geographical spread, and market segments to compete effectively.

In the longer term, a firm needs to focus on the supply side, that is, it needs to develop its *core competences*. These competences are the bundle of skills and technologies that enables the organization to provide a particular benefit or value to customers (Hamel and Prahalad, 1994). This supply-side approach is the so-called resource view of the firm. The firm must know what it is good at or add value to customers in order to succeed. According to Hamel and Prahalad, the firm must have "strategic intent" (p. 141) or "the dream that energizes a company" to "stretch" itself rather than merely "fit" existing resources to emerging opportunities. Thus, strategic intent creates, by design, a substantial "misfit" between resources and aspirations (p. 142).

For a large organization, there are generally three strategic levels, namely, corporate, business, and functional levels. A transnational firm with a corporate head office, divisional businesses, and functional departments within these divisions is a good example of these strategic levels. However, it is often not appreciated that these levels are only conceptual rather than physical, and they need not be physically separated (e.g. head and branch offices).

At the corporate level, the key tasks for senior management are to define the corporate mission (that is, the reason for its existence and portfolio of businesses), vision (what it aspires), and corporate philosophy in terms of

- its relation with other firms, stakeholders, broad objectives (e.g. growth and profitability); and
- values (e.g. innovation, professionalism, and trust).

The management philosophy of Louis Gerstner (2002), ex-CEO of IBM, is a good example. The abridged version is as follows:

- I manage by principle, not procedure.
- The market place dictates everything we should do.

- I'm a believer in quality, strong competitive strategies and plans, teamwork, payoff for performance, and ethical responsibilities.
- I look for people who work to solve problems and help colleagues. I sack politicians.
- I am heavily involved in strategy; the rest is yours to implement...Don't hide bad information – I hate surprises.
- Move fast.
- Hierarchy means very little to me. Let's put together in meetings the people who can help solve a problem, regardless of position.

In developing the corporate mission, the organization needs to determine what businesses it is in, should be in, and should not be in. Examples of mission statements are given below:

"To bring inspiration and innovation to every athlete in the world." (Nike)

"People working together as one global company for aerospace leadership." (Boeing)

Note that Boeing incorporates its vision (aerospace leadership) into its mission statement. It is a matter of management preference whether the mission and vision should be separated.

How does an organization develop its mission, and why does it matter to have a corporate mission? The answer to the second part is simple: without an appropriate mission, the organization may be competing in the wrong business. To develop a mission, the organization first does a SWOT analysis, that is, it identifies its internal strengths and weaknesses (SW), and external opportunities and threats (OT). All too often, strategic planning rapidly degenerates when an organization exaggerates its strengths, underplays its weaknesses, and wrongly identifies threats and opportunities.

In conducting a SWOT analysis, the organization should examine the following factors:

- leadership;
- corporate culture;
- incentive structure;
- inputs, suppliers, and logistics;
- processes;
- pricing;
- quality;
- development time;
- service;
- sales network;
- skills and experience;
- financial position;
- existing and new markets;
- cyclical factors;
- competitors;

- branding and publicity;
- certification;
- location;
- legal, administrative, and information technology infrastructure; and
- changes in legislation.

Next, the organization develops its strategic posture by identifying several strategic *thrusts* to improve or develop products, services, and markets. For instance, the vision of the Singapore Government with respect to information and communication technologies (ICT) is to be a leading e-Government to better serve the nation in the digital economy. The five strategic thrusts are to

- push the envelope of electronic service delivery;
- build new capability and new capacity;
- innovate with ICT;
- be proactive and responsive; and
- develop thought leadership on e-Government, that is, to sensitize public servants to the impact of ICT (Tan, 2000).

For comparison, we provide Asia Development Bank's (2003) strategic thrusts for e-development of ICT in Asia and the Pacific:

- to create an enabling environment by fostering the development of innovative sector policies, strengthening public institutions, and development of ICT facilities, related infrastructure, and networks;
- to build human resources to improve knowledge and skills and promote ICT literacy and lifelong learning through e-learning and awareness programs; and
- to develop ICT applications and information content for ADB-supported activities, e.g., poverty reduction and good governance.

As a final example, the strategic thrusts for a private organization may consist of

- developing core capabilities through staff development, training, and new hires;
- building two new plants in X and Y;
- consolidating existing markets through aggressive marketing, pricing, and new sources of supply; and
- developing new markets through strategic alliances and partnerships.

Finally, top management allocates resources to business units for major projects in each industry, sector or division. Financial considerations, particularly funding, profitability, growth, and cash flows, play major roles here. Many business strategies are executed through projects.

The corporate headquarters also develops key human resources, its core competences, and establishes an administrative infrastructure for the entire organization.

Corporate headquarters vary substantially in sizes depending on their priorities on these functions. It is also possible that the corporate headquarters is bloated, particularly if it is obsessed with planning function.

3.4 Business strategy

A strategic business unit (SBU) or simply "business unit" is a division of the organization. For a university, its business units are the colleges or institutes. In a large firm, it could be a product or geographic division (e.g. coffee or Europe division).

At the business level, the strategic concern is how to effectively compete in the industry. It therefore differs from that of the corporate headquarters discussed above.

How does a business unit compete effectively? The answers are varied, but it first does an internal (efficiency) scrutiny of its value chain (Porter, 1985) in terms of its

- primarily activities comprising inbound logistics, operations, outbound logistics, marketing, and service, and
- its supporting business infrastructure, technological development, and procurement.

Secondly, the business unit does an external analysis of the industry it is competing. Porter's (1980) 5-Forces model is widely used here. The intensity of rivalry in an industry depends on

- the number of competitors;
- bargaining power of buyers;
- bargaining power of suppliers;
- the cost and availability of substitutes; and
- the threat of new entrants.

This largely neoclassical view of competition may be contrasted with the Marxian view that competition extends to all factors of production including competition for capital, land, and workers as well as competition among industries beyond substitutes to equalize the rate of profit. Hence, in the Marxian view, every industry competes for the consumer dollar.

Given its value chain and the intensity of rivalry in an industry, the business unit may compete

- on cost in industries where barriers to entry are low (e.g. construction industry) or where products are homogeneous (i.e. commodities);
- on quick and efficient service with reasonable pricing (e.g. McDonalds);
- on quality through branding or innovation (i.e. product leadership);
- on a particular segment of the market that is largely untapped (e.g. long-distance learning programmes that are not offered by elite universities);
- by providing total customer solutions (e.g. IBM); and

- by locking-in customers by establishing an industry standard (e.g. Microsoft's Windows operating system).

A business unit may compete on more than one front, such as in the airline industry where established airlines compete among themselves as well as with budget airlines. However, it may lose its focus (Kaplan and Norton, 2001). Market intelligence in spotting trends and segment gaps are clearly important in deciding where and how to compete.

Once a business unit has decided where to position itself, it develops action programs to execute its strategy (Bossidy et al., 2002). The so-called "key success factors" that have been actively promoted in the popular business management literature include

- a balanced scorecard of financial and non-financial performance measures (Kaplan and Norton, 1996) that include benchmarking, customer service, training, internal processes, and adoption of best practices;
- a "strategy map" (Kaplan and Norton, 2004) that describes the corporate strategy through cause and effect relationships and how it achieves its scorecard;
- strong leadership to provide the vision, motivate, and reduce resistance to new ideas (Kouzes and Posner, 1995);
- a simple structure for decision-making, accountability, coordination, and information sharing (Mintzberg, 1993);
- efficient processes to manage operations, customers, innovation, and regulatory and social processes (Kaplan and Norton, 2004; Hammer and Champy, 1993);
- core competences and innovation (Hamel and Prahalad, 1994) rather than a portfolio of unrelated production units (i.e. a conglomerate);
- a strong culture of creativity, achievement, discipline, and ownership (Deal and Kennedy, 1982); and
- incentives and control to align effort towards achieving strategic goals (Manas and Graham, 2003).

A balanced scorecard is required to go beyond purely financial performance measures. It includes non-financial indicators such as number of defects, delays, customer satisfaction, training, and learning. The underlying message with performance indicators is the common echo that "you cannot manage or improve on what you cannot measure."

A strategy map describes the corporate strategy in the form of cause and effect relationships. In essence, "you cannot manage what you cannot describe" as well. The strategy needs to be articulated clearly. Figure 3.2 shows the main components of a strategy map. It shows what needs to be considered (within the boxes) and their relations (as shown by arrows).

While strong leadership is necessary if the organization is to have a vision and execute it flawlessly, leaders are not necessarily charismatic. They are more typically the builders of the organization and shapers of corporate culture (Collins and Porras, 1994).

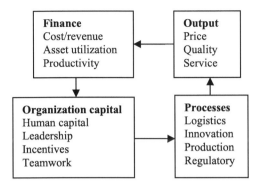

Figure 3.2 Elements of a strategy map.

An organization structure is required for decision-making, accountability, and information sharing.

Traditionally, organizations were organized along functional lines (e.g. financial, human resource, legal, information technology, production, sales, and so on). Such a structure tends to create functional specialization and turf (boundary) problems because of suboptimal departmental goals.

As the products and services of conglomerates proliferated, autonomous divisions based on geography or products were set up after World War II. The downside of having autonomous divisions is some duplication of functions but this is a small price to pay for greater accountability of divisions as profit centers rather than cost centers. Within each division, functional departments were set up and, to overcome departmental politics, shifting of blame for product failures, and slow product development, many matrix structures based on project teams and led by project managers (PMs) were set up (Figure 3.3). Such project structures generally work well if the conflict between functional and project departments can be sorted out. It is possible that functional managers who control functional resources may jeopardize a project by under-allocating these resources.

By the end of 1980s, many Western conglomerates found that they could not compete effectively in too many product markets. Until then, the dominant thinking was to have a portfolio of diversified and uncorrelated businesses to lower the overall portfolio risk. If a few product lines or regions fail, the conglomerate can always rely on the remaining successful ones to stay in business.

The initial response to the inability to compete effectively in too many markets was to cut cost and improve productivity and quality. Soon, it was realized that delayed, downsized, or "right-sized" firms were already cut to the bone, and there was something more than cost-cutting and endless rhetoric on quality and productivity.

Such a firm was subsequently said to lack strategic focus. One common solution was to switch tactics and divest unrelated product lines and focus on the firm's core competences.

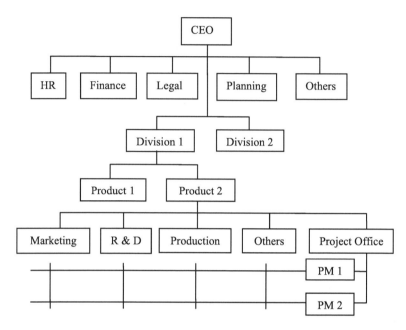

Figure 3.3 Organization structure for a large corporation.

In the 1990s, many corporations also reengineered their operations, customers, innovation, and regulatory and social processes using project teams to cut R & D and production time as well as provide better services to customers. Reengineering entailed a radical redesign or reinvention of work processes. However, the reengineering movement tended to forget that production is more than a technical process. It is also a social process involving cooperative workers, and there were just as many reengineering failures as well as successes.

The success of Japanese firms in the 1970s, 1980s, and 1990s led many commentators to speculate that, beyond just fairly technical reengineering, a strong corporate culture seems to matter. A consistent set of beliefs, values, and practices was thought to drive management and workers towards achieving the corporate mission and vision in a harmonious way. In the Japanese case, a culture of trust, loyalty, care for workers and their families, and practices such as lifetime employment and job rotation were contrasted with the alleged Western "distrust" between management, unions, and government, the considerable use of short-term employment contracts, lack of identification with the company, and excessive specialization.

A further issue was how to align incentives and develop controls to raise productivity and improve quality. Incentives are viewed as a total package that includes salary, promotion, perks, training, mentoring, time-off, and recognition. In theory, these incentives should be carefully evaluated and linked to performance. In practice, performance appraisal was both time-consuming and difficult to implement fairly.

Although different authors tend to stress different factors, these factors must generally "fit" (or align) as a package for the organization to succeed (Miles and Snow, 1994). It was noted earlier that an organization should "stretch" itself to realize its aspirations.

3.5 Functional strategy

Recall from Figure 3.1 that even a matrix organization contains functional departments. The key functional strategy is to enhance functional capabilities to support corporate and business (divisional) strategies. The process must be properly managed to reduce waste and improve efficiency by identifying value-creating activities along a value stream and follow it with flawless execution. The tools include

- continuous improvement or "Kaizen" including the Deming cycle (Imai, 1991);
- business process reengineering (Hammer and Champy, 1993);
- lean production (Womack and Jones, 1998);
- total quality management (Evans and Dean, 2003);
- supply chain management (Burt et al., 2003); and
- Six Sigma (Pande et al., 2000).

These ideas (which sometimes conflict, for example, continuous process improvement and radical process reengineering) were consolidated and expanded in Six Sigma. The additional elements (though hardly new) in Six Sigma include customer focus, fact-driven management, proactive management, and boundaryless collaboration (teamwork) within the firm and with suppliers and customers. Fact-driven management is derived from the desire to control statistical variation tolerance to within six standard deviations, that is, almost defect-free. If $f(x)$ is the probability density function of a standard normal variate, then

$$\int_{-3}^{3} f(x)dx = 0.999997 .$$

The quality control in Six Sigma is a stringent criterion. It is an ideal rather than a key performance indicator.

3.6 Strategic project office

As we have seen, a major weakness of the functional structure is the absence of ownership of the entire process as activities are passed from one functional department to another. This results in problems such as

- the tendency to "pass the buck" to another department when things go wrong;
- delays in bringing products to market because any department can hold up the entire process; and
- poor coordination.

Hence, it is possible that the production department may refuse to manufacture the product because of "technical difficulties." This problem would have been avoided if inputs from suppliers, production, and marketing were also used during the design stage.

For these reasons, cross-functional project teams, ad hoc committees, concurrent engineering, and reengineering were used to integrate processes across functional departments. Project managers reside in the (physical or virtual) Strategic Project Office (SPO) or Enterprise Project Management Office (EPMO) headed by the Director of Projects to link corporate strategy and seamless project execution.

Such a setup encourages the sharing of expertise and use of templates, standards, best practices, and common tools to effectively select and deliver projects. Hence, a key task of an SPO is to develop project management competency using an organizational project management maturity model (OPMMM or OPM3), and this is briefly discussed below.

3.7 Project management maturity

Current organization project management maturity models tend to draw heavily from the Software Engineering Institute's (SEI) Capability Maturity Model (CMM). Developed in 1986 for IT projects (SEI, 1995; Humphrey, 1989), the descriptive CMM reference model contains five levels of maturity, namely,

- an initial process without established practices;
- basic documentation of separate processes;
- documentation and institutionalization of entire project management process;
- application of project management processes across all projects and quantification; and
- optimization of processes through deliberate project feedback, learning, and improvement.

The Project Management Institute's (PMI) Organization Project Management Maturity Model (OPM3) uses the same five maturity levels across nine project management knowledge areas, namely,

- project integration;
- scope management;
- time management;
- cost management;

- quality management;
- human resource management;
- communications management;
- risk management; and
- procurement management.

For instance, under risk management, maturity proceeds as follows:

- level 1: no established risk management tools;
- level 2: basic documentation of risk management framework;
- level 3: institutionalization and integration of risk management framework within the organization;
- level 4: application of risk management framework across all projects; and
- level 5: optimization and improvement of risk management framework.

Organizations can therefore improve their project management maturity level by level.

An organization needs to assess its project management maturity. This may be done internally or through independently certified assessors or consultants. There are pros and cons of using each approach. Internal assessors know the internal politics and workings of an organization better. However, external assessors tend to be more objective and may be able to introduce best practices for the organization to adopt.

3.8 Public organizations

The State is generally regarded as a set of institutions (i.e. organizations) that possess the authority to make rules that govern a territory or society. The key institutions of a modern State are

- the legislature (e.g. parliament) to make rules;
- the judiciary (courts) to interpret these rules;
- the executive (bureaucracy) to implement them;
- the police to enforce the laws; and
- the military for territorial defense.

Within a federal system, political power is also shared among the federal, state, and local governments. Further, there has been vertical re-scaling of governmental boundaries from local to international treaty organizations (e.g. the European Union) and possible "scale jumping" in dealings among different levels of governments. These developments have both consolidated and fragmented State power depending on the contextual constellation of political forces. For instance, some governments have consolidated power by removing government at the metropolitan scale. On the other hand, other national governments such as those in the European Union have seen their powers eroded by increasing economic and political integration.

The conceptual boundary of the State is debatable. In some conceptions, political parties, religious institutions, the mass media, and educational institutions are also considered part of the "ideological apparatuses" of the State. Although the State has monopoly of force within its territory, Gramsci (1971) argued that it is seldom used to legitimize its rule. Rather, economic performance and hegemonic ideology (i.e. accepted ruling ideas) are more important means for politicians to get elected or stay in office.

The State bureaucracy or civil service is divided into ministries answerable to the elected ministers. In turn, politicians are answerable to voters. Since ideologies clash, it is often the case that power is shared among political parties and this can lead to a bloated bureaucracy, corruption, turf politics, overblown budgets, and difficulties in coordinating the various agencies. Similarly, changes in political power can lead to oscillating ideologies and policies.

Traditionally, the problems besetting the civil service were viewed as largely "administrative" problems and hence dealt with administratively by measures such as

- setting up a public service commission to deal with promotions of senior civil servants across ministries;
- setting up a corrupt practices investigation bureau;
- breaking up large ministries into autonomous statutory boards subjected to parliamentary controls on budgets; and
- rotating senior public servants to prevent "empire-building."

It is not surprising that, during the 1950s and 1960s, civil servants were also commonly called "public administrators" working in the "administrative service" and universities offered programs in "public administration." From the 1970s, private sector management practices were increasingly implemented in the public sector. The plethora of tools included management by objectives, computerization, quality management, performance appraisal, faster promotion, salary bench marking, and flexible salary structures.

By the end of the 1980s, many public sector organizations were developing corporate strategies and road maps found in private firms. They were also talking about corporate culture. Many unprofitable State-owned enterprises (SOEs) were corporatized or closed down, and units with profit potential but no strategic significance were privatized or publicly listed. Monopolized industries were also deregulated to make firms in these industries more competitive by leveling the playing field. New regulatory agencies were set up to become enablers or facilitators rather than service providers. Lastly, public-private partnerships were formed to help cash-trapped governments develop the much needed infrastructure projects.

The current roles of the State in a changing world are neatly summarized in two World Bank (1997; 2002) Reports. According to these reports, the modern State should

- focus its activities to match its capabilities and try not to do too much with too few resources and little capability;

- improve its capability by reinvigorating public institutions to provide civil servants the incentives to do their jobs better, be more flexible, and also restrain arbitrary and corrupt behavior; and
- build institutions for markets.

These points are significant. The first point suggests that States (i.e. politicians) should not make too many empty promises when running a State with few resources and limited capabilities. Second, much can be done to provide incentives for civil servants to do their jobs better. Thirdly, the State should build institutions to complement the market. The old debate between State or market is replaced by a new question on how the State can act (i.e. design new rules or institutions) to make markets work better. Institutions can support markets by

- defining and enforcing property rights;
- reducing the transaction costs of: (a) searching for price and quality information, (b) writing, concluding, and enforcing contracts, and (c) monitoring opportunistic behavior;
- aligning incentives; and
- increasing competition.

Building institutions begins with identifying the property rights, transaction costs, incentives, and level of competition in each sector. The design of institutions then proceeds by complementing existing institutions and identifying new institutions that work.

Questions

1   It is often said that a leader does the right thing and a manager does things right (Drucker, 1967). Explain why this may not be an accurate view of what managers do.

2   In the training of project managers, what competencies are required, and why are they important?

3   This is how ex-CEO of IBM, Louis Gerstner (2002: 46), described his first strategy conference at IBM as CEO in 1993:

> The presentations were both formal and formidable. I was totally exhausted at the end.... The technical jargon, the abbreviations, and the arcane terminology were by themselves enough to wear anyone down.... There was little true strategic underpinning for the strategies discussed. Not once was the question of customer segmentation raised. Rarely did we compare our offerings to those of our competitors.

What are the missing strategic elements?

4   This is what Gerstner (2002: 47) found at a customer meeting:

On Tuesday night, I met with several Chief Information Officers at dinner .... They were angry at IBM – perturbed that we had let the myth that "the mainframe was dead" growth and prosper. The PC bigots had convinced the media that the world's great IT infrastructure – the back offices that ran banks, airlines, utilities, and the like – could somehow be moved to desktop computers. These CIOs knew this line of thinking wasn't true, and they were angry at IBM for not defending their position. They were upset about some other things, too, like mainframe pricing .... They were irritated by the bureaucracy at IBM and by how difficult it was to get integration.

While this passage clearly underscores the importance of customer focus, why is it that many firms did little but merely "talk" about listening to customers? What does it mean to have a customer focus?

5   According to Imai (1991), Kaizen means *continuous* process improvement involving everyone. However, Hammer and Champy (1993) argued that the key idea in business process engineering is nothing less than a *radical* reinvention of how they do their work. Reconcile the two seemingly contradictory approaches.

6   According to Porter (1980), firms can use three generic strategies. Apart from a niche or focus strategy serving a small segment of the market, firms that compete broadly in scope should pursue either a low cost strategy or a differentiated strategy based on quality. A strategy that produces a medium cost, medium quality product is "stuck in the middle." Explain why the concept of "medium quality" is problematic in this classification of generic competitive strategies.

7   "Corporate culture is little more than motherhood statements on innovation, openness, teamwork, and the like. They are part of rah-rah management to be forgotten not long after the party or meeting is over. It is indeed insulting to subject intelligent employees to this kind of silly corporate game." Does corporate culture really matter, and why?

8   According to O'Connor (1973), the capitalist State needs to perform two basic functions: to assist accumulation of capital, and to legitimize its rule. In performing the accumulation function, the State needs to enhance private profitability by investing in unprofitable infrastructure and other social capital expenditures (e.g. housing, medical, and education expenditure). In performing the legitimization function, the State needs to incur welfare and warfare expenditure (e.g. policing) to secure social harmony. As a result, the State faces a structural fiscal crisis. Explain why this thesis runs contrary to evidence on State budget surpluses.

9   The financial system in a less developed country is often said to be "weak" or "undeveloped," meaning that, among other things, it is fragile, borrowing costs and credit risks are high, property rights (e.g. titles to property to be used in

mortgages) are not secured, markets are missing (e.g. derivative markets), and information are not readily available. How can the State strengthen such a financial system?

4

Corporate Finance I

4.1 The balance sheet

Corporate finance deals with the "financing of projects" based on full recourse lending. That is, lenders have full recourse to a firm's assets should the project fail, unlike non-recourse or limited-recourse "project financing" where lenders rely primarily on the (tangible and intangible) assets and cash flows of the project to service the debt.

In corporate finance, two statements are of considerable importance and extensively studied by investors, namely, the balance sheet and the income statement.

The balance sheet provides a picture ("snapshot") of the financial posture of the firm at a particular date (e.g. 31 Dec 2005). Despite certain shortcomings, financial statements provide valuable information on corporate finance. Such statements are based on Generally Accepted Accounting Principles (GAAP), of which two are basic:

- accrual accounting basis where revenue from selling a good or service is recognized in the period in which the good is sold or the service is substantially performed irrespectively of whether the firm is paid during the period; and
- the accrual principle also applies to the expense side to match expenses to revenues.

In short, a firm apportions revenues and expenses to a particular period of time even though it has not received payment from its customer for its sales or paid a supplier for inputs. Accrual accounting better reflects the firm's actual financial performance.

Expenses are categorized into

- operating expenses in the current period (e.g. labor, material, marketing, and administrative costs);
- financial expenses on debt (e.g. interest); and
- capital expenses on fixed assets that are used to generate revenue beyond the current period (e.g. land, building, and equipment).

The balance sheet has three major items (Table 4.1), namely,

- assets;
- liabilities; and
- net worth (i.e. assets minus liabilities), also called shareholders' equity, which is positive if assets exceed liabilities, and negative otherwise.

Assets ($m)		Liabilities ($m)	
Current assets		**Current liabilities**	
Cash and Treasury bills	20	Accounts payable	14
Accounts receivable	50	Dividend and taxes payable	6
Inventories	40	Short-term loans	10
Total current assets	**110**	Total current liabilities	**30**
Long-term assets		**Long-term liabilities**	
Land and buildings	50	Long-term bank loans	20
Machinery and equipment	30	Bonds	30
Depreciation	(10)*		
Total long-term assets	**70**	Total long-term liabilities	**50**
Intangible assets			
Patents and copyrights	**10**		
Total assets	**190**	Total liabilities	**80**
		Net worth (Shareholders' equity)	**110**
		Common stock	60
		Retained earnings	50
		Liabilities + net worth	**190**

Table 4.1 The balance sheet.

* (.) denotes a negative figure by accounting convention.

A firm's assets comprise

- current assets that are liquid (cash and deposits, T-bills, and accounts receivable) and inventories of raw materials, work-in-progress, and unsold finished goods;
- long-term assets (land, buildings, vehicles, and machinery); and
- intangible assets (e.g. patents, trademarks, goodwill, trade secrets, and copyrights).

Other types of the firm's "assets" such as management systems, established procedures, human capital, and relations with suppliers, partners, stakeholders, and customers (or "relational capital") do not appear in the balance sheet. With the exception of research and development which is reported as an operating expense rather than more appropriately as capital expenditure, this incomplete picture in the balance sheet makes it difficult to value internet and technology firms (Damodaran, 2001; Kettell, 2002).

Traditionally, one adds an additional 10 per cent to asset book value for such intangibles. However, in a knowledge economy, this procedure grossly understates the value of intangible assets. For instance, the average ratio of the market value of the firm based on stock prices to net assets of S & P 500 firms has risen from about 1.0 in the early 1980s to 6.0 in 2001 (Lev, 2001). Nonetheless, the values of intangible assets are difficult to ascertain econometrically because of recurring misspecification and measurement problems (Griliches, 1977; Nadiri, 1993; Jones and Williams, 1998).

A firm's liabilities include

- current liabilities (accounts payable to suppliers, dividend payable, taxes payable, and short-term loans);
- long-term loans; and
- value of outstanding bonds.

In Table 4.1, the firm has a net worth of $110 m comprising $60 m of common stock (i.e. money initially invested in the company by investors who are shareholders) and $50m of retained earnings. Since

Assets = Liabilities + Net Worth,

the balance sheet must, apart from rounding errors, balance.

4.2 Analyses of balance sheets

What do investors and lenders look for in a firm's balance sheet before they decide to invest, buy over, or lend to the firm?

The current assets show part of a firm's liquidity position. The other part is revealed by its liabilities. On the assets side, if it is cash rich, the firm may simply be conservative or it may be looking for new investment opportunities or acquisitions in the near future. Alternatively, the firm may buy back its shares (thereby raising its share price) as part of its capital reduction plan or declare higher dividends to reward shareholders.

A high level of inventory is costly in terms of opportunity and carrying costs (i.e. insurance, storage, damage, and risk of obsolescence). If the firm is not stuck with a high level of unsold goods because of poor sales, it may be inefficient in inventory management. Inventory carrying costs can be high and, for this reason, "zero inventory" just-in-time (rather than "just-in-case") production systems such as the Toyota system are popular.

For long-term assets, firms are required to periodically revalue their land and buildings and reflect the fair market values in the balance sheet. Since property values can rise or fall within an accounting period and, to the extent that they do not affect operations, the change in value is merely paper gain or loss. From a different perspective, property values can change substantially over time and this will affect the value of a firm's assets. In turn, changes in asset values will affect a firm's ability to borrow and hence its operations.

Unlike property, the values of machinery and equipment are often booked at historic cost and seldom revalued upwards because of their values generally do not appreciate over time. Rather, depreciation charges are applied to historic costs in revaluing these assets.

The amount of depreciation allowed for buildings and equipment depends on the tax code. Generally, land cannot be depreciated. Depreciation for buildings depends on the type of building (e.g. 27.5 years for residential buildings, and 39 years for commercial buildings). Computers are normally depreciated over five years, vehicles are depreciated over 10 years, and so on.

Suppose a machine is bought for $50,000 and has a scrap value of $5,000. If the allowable depreciation is 10 years, the annual depreciation using the straight line method is

$$(50,000 - 5,000)/10 = \$4,500.$$

In contrast, if depreciation is accelerated to 5 years, the annual depreciation is

$$(50,000 - 5,000)/5 = \$9,000.$$

Double declining-balance depreciation is another form of accelerated depreciation. If the allowable depreciation period is 10 years, then

$$\text{Depreciation expense} = \text{Depreciated value} \times 2/10$$

The word "double" comes from the factor of two in the numerator. Using the same machine example, for the first year,

$$\text{Depreciation expense} = \$50,000 \times 2/10 = \$10,000.$$

For the second year, the depreciated value (or net book value) is $50,000 - $10,000 = $40,000. Hence,

$$\text{Depreciation expense} = \$40,000 \times 2/10 = \$8,000,$$

and so on.

Improvements are depreciated at the same rate as the original asset. Like the revaluation of land values, the annual allowable depreciation may not reflect the productive or economic life of the machine. For instance, major airlines used to lengthen the depreciation period for planes to book lower depreciation expenses and hence report higher earnings.

4.3 Financial ratios

Financial ratios are rough measures of a firm's financial health, particularly its liquidity and ability to service its debt. They are "rough" because such ratios are merely rules of thumb, just like the way ratios are used to describe objects or patterns (e.g. ratio of length to breadth).

A ratio can also be converted into an index. For instance, the well-known body mass index (BMI) is defined as the ratio of an adult person's weight in kilograms divided by the square of her height in meters. It is used to study obesity where BMI exceeds 30.

Obviously, many financial ratios can be computed by using various combinations of variables, and only the basic ratios are given below:

Current ratio = Current assets/Current liabilities
= 110/30 = 3.67.

Net working capital = Current assets – Current liabilities
= 110 – 30 = $80 m.

Liquidity ratio = net working capital/total assets
= 80/190 = 0.42.

Debt/equity ratio = 50/110 = 0.45.

Net tangible assets (NTA) = total assets – intangible assets – liabilities
= 190 – 10 – 80 = $100 m.
= Net asset value (NAV)

Financial ratios are historical and may not a firm's future or potential financial position.

A firm manages its cash to retain sufficient liquidity and invests excess cash in short-term T-bills and bank deposits. Obviously, accurate forecasts of future cash inflows and outflows are important. The firm has to manage its accounts payable and accounts receivable in terms of how much credit to extend to customers and obtain from suppliers, billing policy, and debt collection. In the short term, shortfalls may be met, if necessary, with bank overdrafts and lines of credit.

4.4 Off-balance sheet items

A firm may use off-balance sheet financing to keep its debt off its balance sheet so that its debt ratio is kept low. Generally, debt covenants imposed by lenders require low debt ratios, but another motive for off-balance sheet financing of a capital-intensive project through a separate entity or special purpose vehicle (SPV) is to calm equity investors to support its share price.

Examples of off-balance-sheet financing include joint ventures, partnerships, loans to separate entities guaranteed by the parent, and operating leases. In the interest of greater transparency, listed firms are often required to disclose off-balance sheet items in their annual reports (or at least in the footnotes). Without off-balance sheet financing, project sponsors may not be able to undertake the risks inherent in large and complex projects.

Off-balance sheet financing may be abused. For instance, Enron created SPVs to hide its huge debt before it collapsed in 2001 (Fox, 2003). Formed in 1986 through a merger of two pipeline companies, Enron set up subsidiaries as follows (Figure 4.1):

- Enron Online to tap the potential in the Internet by creating an e-commerce website in 1999 to trade in commodities and financial instruments;
- Enron Capital and Trade Resources (ECT) to benefit from global trading in deregulated energy markets;
- Enron International for international operations; and
- Enron Finance Corporation (EFC) to carry out the financing of energy projects.

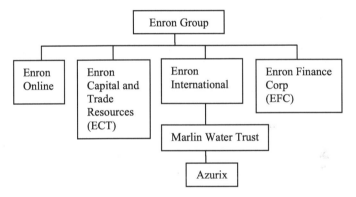

Figure 4.1 Elements of the Enron Group.

As a fee-based entity and early starter (or innovator) of the dot.com boom, Enron Online was profitable even up to the time Enron filed for bankruptcy in 2001.

ECT was entirely different. Its traders bought companies that did not fit Enron's core energy business. These activities of these companies included paper manufacturing, steel-making, and providing Internet services. Even when ECT did acquire energy businesses, such as its acquisition of deep sea oil and gas explorer Mariner Energy, critics argued that its value is overstated. Deep sea oil exploration is risky, and a little-known firm such as Mariner Energy can be worth millions or little depending on how its potential is valued. Apparently, this vagueness is sometimes exploited by rogue traders, including those at ECT. Of course, there is nothing wrong with running a risky business. However, deliberately overstating a firm's value crosses the line.

The next company in the Enron stable, Enron International, incurred huge debts from many sources early in the game. There were expensive acquisitions such as the failed US$2.8 bn Dabhol power project in India in the early to mid-1990s, partly because corporate bonuses were paid to "bright and talented star players" to close deals. The project was stalled by protests over environmental concerns, the displacement of eight villages, and the pricing of electricity. Since the long-term off-take (output) contract was denominated in US dollars, if the greenback appreciated against the rupee, Indian consumers would need to pay more and more for power. In 1996, the Indian Congress Party was voted out of office, and the new government stopped the project.

In 1998, Enron International set up a special purpose vehicle, Marlin Water Trust, to finance its US$2.2 bn purchase of Wessex Water and renamed it Azurix. To keep the purchase off its books, Enron International managed to find investors to take a 50 per cent stake in Marlin but Azurix was subsequently floated in 1999 with Enron retaining a 35 per cent share. From the start, Azurix was doomed to fail. Water was not Enron's core business, and it did not fully understand that the water industry was tightly regulated in many countries for political reasons and resulted in low margins. In Argentina, it bid US$438 m for a 30-year concession to supply water, much higher than any other bidder.

Many other SPVs were set up by Enron (not shown in Figure 4.1) to hide its debt (see McLean and Elkind, 2003; Eichenwald, 2005).

4.5 The income statement

A firm's income statement provides a picture of the profit or loss made during the year (Table 4.2). Hence, it is also called a profit and loss statement.

The firm has sales revenue of $100 m. The cost of goods sold refers to the *direct* costs incurred in producing the goods or services. It includes the cost of raw materials, energy, and direct labor. After deducting the cost of goods from the sales revenue, we add the change in inventory (which may be positive or negative), which is zero in Table 4.2. The firm earns a gross profit of $60 m. On the other hand, if the inventory is $5m on 1 January 2005 and $6 m on 31 Dec 2005, the change in inventory is not zero but −$1 m.

The operating income (or operating profit) of $25 m is obtained by deducting from gross profit the general operating expenses. These *indirect* expenses are not attributable to a particular item for sale. They include research and development expenses, insurance, accounting, rent, marketing, utilities, shipping, salaries, printing and other administrative expenses, and depreciation.

Other income such as income from investments and extraordinary income are then added (or deducted if a loss is reported) to operating income to obtain the earnings before interest and tax (EBIT) of $30 m. EBIT is important because it shows the ability of the firm to pay its debt.

Interest expense and corporate tax are then deducted from EBIT to derive the net earnings to be distributed as dividends or retained within the firm.

Item	2005 ($m)
Sales revenue	100
Less: Cost of goods sold	(40)
Change in inventory	0
Gross profit	**60**
Less: General operating expenses	(30)
Less: Depreciation expense	(5)
Operating income	**25**
Other income	5
Earnings Before Interest and Tax (EBIT)	**30**
Less: Interest expense	(5)
Less: Taxes	(10)
Net earnings	**15**
Less dividends	(5)
Addition to retained earnings	**10**

Table 4.2 Income statement.

4.6 Operating ratios

What do lenders and investors look for in an income statement? The income statement should be analyzed over a number of years and across firms in the same industry to benchmark and identify trends in each category. One can then determine if sales, costs, and profits are growing, stagnating, or falling and if they are above or below the industry average.

Just as financial ratios are used to determine the financial health of the firm from its balance sheet, many operating ratios may be computed from the income statement to determine operating efficiency. The basic ratios include

- Cost of goods/Sales revenue;
- General operating expenses/Sales revenue; and
- Operating expense/Operating income.

It is possible to analyze the breakdown of each category. For instance, if cost of goods has risen, one may wish to know whether it is due to rising raw materials, labor, or other costs.

4.7 Profitability ratios

Many ratios have been suggested as rough measures of profitability. The following ratios are often used:

$$\text{Return on assets (ROA)} = (\text{EBIT} - \text{Tax})/\text{Total assets}$$
$$= (30 - 10)/190 = 10.5\%.$$

$$\text{Return on equity (ROE)} = \text{Net earnings/equity}$$
$$= 15/110 = 13.6\%.$$

For many industries, ROA and ROE range between 10 to 20 per cent. The differences in returns reflect the risks and level of competition in each industry. The latter affects the ease of entry of new firms.

4.8 Free cash flow

The amount of free cash flow (FCF) in a firm is given by

$$\text{FCF} = \text{Net earnings} + \text{Depreciation expense} + \text{Net working capital}$$
$$- \text{capital expenditure} - \text{dividend}$$
$$= 15 + 5 + 80 - 80 - 1 = \$19 \text{ m}.$$

It is assumed that divided payout is $1m. Note depreciation expense is a non-cash charge, that is, the firm is not actually spending money but imposes the charge for tax purposes.

FCF is a measure of the cash the firm has after it has paid its expenses, capital expenditure (in land, buildings, vehicles, and equipment), and dividend. Some analysts argue it provides a truer picture of a firm's cash position than net earnings. It allows a firm to exploit opportunities for growth.

However, negative FCF is not necessarily a bad thing. The firm may have recently incurred large capital expenditures to boost growth and profitability. Finally, since capital expenditures tend to be "lumpy," FCF may fluctuate considerably from year to year.

4.9 Market ratios

Market ratios relate the firm's market value measured by its share price to some accounting variables. These ratios provide an insight on how well the market perceives the firm is doing.

The most common ratio is the price/earnings (PE) ratio given by

$$\text{P/E ratio} = \text{Price per share/Earnings per share}$$

For example, if the price per share is currently $0.70 and the net earnings per share is 7 cents, then

   P/E ratio = 70/7 = 10.

This ratio may then be compared to P/E ratios of similar firms in the same industry. Firms with high P/E ratios are viewed as relatively more "expensive," but it is important to remind ourselves that it is based on historical data and not future performance.

The Market/Book (M/B) ratio is given by

   M/B ratio = P/V

where P is price per share of common stock, and V is book value per share of common stock. The latter is computed using

   V = Common stock equity/Number of shares of common stock outstanding

For example, from Table 4.1, the common stock equity is $60 m, and if there are 100 m outstanding shares of common stock, then

   V = (60 m)/(100 m) = $0.60.

If the price per share of common stock is $0.70, then

   M/B ratio = 0.70/0.60 = 1.17.

This means that investors are currently paying $1.17 for each $1 of book value of the firm's stock. A high M/B value may signal that investors view the firm's prospects as good.

Questions

1   Based on the accrual accounting principle, revenue is recognized in the period of sales irrespective of whether the firm is paid in the same period or next period. Explain why accrual accounting is desirable in recognizing sales from residential projects under construction.

2   Explain why accrual accounting may also lead to undesirable inflation of revenues in residential projects under construction.

3   Materials purchased in a period are carried over to the next period as inventory if it is not used up. If steel is bought at $x per ton and carried over to the next period where the market price has risen to $y per ton, how should this inventory be valued?

4   Many contractors lease their equipment. Should leases be booked as operating or capital expenses, and why?

5   To appreciate some of the difficulties in measuring the return to human capital, if y is earnings without education, then a person's earnings after s years of schooling is

$$Y = y(1 + \lambda)^s$$

where λ is the constant rate of return to education. Hence,

$$\log(Y) = \log(y) + s\log(1 + \lambda) = \alpha + \beta s$$

If y is assumed to be constant, then $\alpha = \log(y)$ is a constant, as is $\beta = \log(1 + \lambda)$. All else equal, males and females may have different earnings and this is captured using a dummy variable G (1 = male; 0 = female) so that

$$\log(Y) = \alpha + \beta s + \phi G + \varepsilon$$

where ϕ is a parameter and ε is the error term. Additional dummy variables may be added to capture earnings differentials due to nationality, race, and religion. Since working experience (X, in years) is important, and tends to peak during mid-life (i.e. quadratic),

$$\log(Y) = \alpha + \beta s + \phi G + \gamma X - \theta X^2 + \varepsilon.$$

Workers have different abilities, and this may be proxied using class grades (C) so that

$$\log(Y) = \alpha + \beta s + \phi G + \gamma X - \theta X^2 + \omega C + \varepsilon.$$

Some variables may also interact. For instance, the impact of working experience on earnings may depend on gender as well so that it is appropriate to add a new interacting variable, the product of X and G:

$$\log(Y) = \alpha + \beta s + \phi G + \gamma X - \theta X^2 + \omega C + \tau XG + \varepsilon.$$

This econometric model to estimate the rate of return to schooling (β) illustrates the difficulties in estimating β. Answer the following:

a)   What are the key assumptions of the model?
b)   Which variables contain serious measurement problems?

6   The table below shows the balance sheet for a transport operator for FY2005.

Assets ($'000)		Liabilities ($'000)	
Current assets		**Current liabilities**	
Cash and deposits	2,133	Accounts payable	81,528
Accounts receivable	325,523	Provisions	247
Inventories	0	Taxes payable	0
Tax recoverable	6,183	Short-term loans	0
Total current assets	**333,839**	Total current liabilities	**81,775**
Long-term assets		**Long-term liabilities**	
Property, plant and equipment	4,259	Long-term loans	300,000
Investments in subsidiaries	333,191	Provisions	634
Depreciation#	0	Deferred tax liabilities	744
Total long-term assets	**337,450**	Total long-term liabilities	**301,378**
Intangible assets			
Patents and copyrights	**0**		
Total assets	**671,289**	Total liabilities	**383,153**
		Net worth	**288,136**
		Common stock	151,862
		Retained earnings	136,274
		Liabilities + net worth	**671,289**

Fixed assets have been fully depreciated.

Compute the following:

a) Current ratio [4.08]
b) Net working capital [$252,064,000]
c) Liquidity ratio [0.375]
d) Debt/equity ratio [1.04]

What can you conclude about the financial posture of the firm?

7   The table below shows the income statement of the same transport operator for
     FY2005.

Item	2005 ($'000)
Revenue	46,920
Staff and related costs	(23,643)
Other operating expenses	(25,721)
Gross profit	**(2,444)**
Depreciation expense	(1,371)
Operating income	**(3,815)**
Interest and investment income	7,627
EBIT	**3,812**
Interest expense	(14,690)
Taxes	0
Net earnings	(10,878)

Compute the following:

a) Return on assets [0.57%]
b) Return on equity [−3.78%]

What can you conclude about its operations?

8   Sometimes, it is argued that Economic Value Added (EVA) provides a better
     picture of earnings than EBIT. It is defined as

        EVA = EBIT − Taxes − Capital charge

For instance, if EBIT is $10 m, taxes are $2 m, and total capital invested is
$100 m at 5 per cent cost of capital, then

        EVA = 10 − 2 − 5 = $3 m.

Discuss.

5

Corporate Finance II

5.1 Sources of funds

The sources of funds for a firm consists of debt and equity supplied by depository institutions (e.g. banks), investors, insurance companies, mutual funds, pension funds, government, and other agencies (e.g. World Bank and export-promotion agencies). Smaller firms may rely on venture capital and credit from friends, relatives, customers, and suppliers (Figure 5.1).

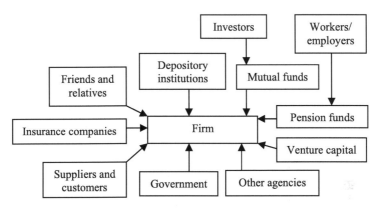

Figure 5.1 Sources of funds.

Each source of funds has its cost and other characteristics such as priority of payment, tax deductibility, and so on (Table 5.1). Established firms tend to fund projects through retained earnings (Atkin and Glen, 1992).

Equity is risk capital, and consists of retained earnings, and funds from issues of preference shares and common shares.

The main sources of debt are loans and bonds. Senior debt has priority in payment over subordinated or junior debt.

Characteristics	Common shareholders	Preferred shareholders	Debt holders
Priority of payment	Lowest	Priority over common shareholders	Priority over shareholders
Tax deductibility	No	No	Yes
Voting rights	Yes	No	No
Cost of capital	Higher than preferred shareholders and debt holders	In between common shareholders and debt holders	Lowest because of tax deductibility and priority of payment
Right to receive assets upon dissolution	Yes	Yes	Full or limited recourse

Table 5.1 Summary of characteristics of debt and equity financing.

5.2 Preferred stock

Preference shareholders have priority over common shareholders in the distribution of earnings and funds in liquidation but do not have voting rights to elect directors and vote on special issues. They are paid a fixed annual (or semi-annual or quarterly) dividend in perpetuity as long as the firm is financially able to do so. Thus, preferred stocks are relatively safe investments but investors have to weigh this against giving up their voting rights and the inability to profit from share appreciation.

Some preferred stocks contain adjustable dividends according to some formula but we shall assume dividends to be fixed. Unlike common stock, the issuance of preferred stock does not dilute share ownership. However, there are preference shares that may be converted to common stock and these are usually callable by the firm, that is, it can force holders of preferred stock to exercise their option to either accept the par value or common shares.

The cost of a simple fixed dividend, non-convertible preferred stock is the value of r_P in

$$V = \frac{d}{1+r_p} + \frac{d}{(1+r_p)^2} + \cdots = \frac{d}{r_p}$$

where V is issue price less flotation cost and d is annual dividend. Hence,

$$r_P = d/V. \tag{5.1}$$

Example

A firm wishes to raise $10 m and issues preferred stock at a par value of $20 per share with a dividend rate of 5 per cent and a floatation cost of $0.10 per share. If similar preferred stocks are currently earning 6 per cent per year,

a) What is the effective cost of preferred stock?
b) How many shares should the firm issue?

a) The effective cost of preferred stock is

$$r_P = d/V = 0.05(20)/19.90 = 5.025\%.$$

b) If the rate of return required by investors is 6 per cent per year, then there will not be any subscription at $20 per share. Instead, the firm must price the share at

$$V = d/r_p = 0.05(20)/0.06 = \$16.67 \text{ per share.}$$

The firm must issue ($10 m)/$16.67 = 599,880 shares.

Preferred stock is generally more expensive than common stock because of the absence of tax deductibility and dividends must be paid as long as the firm is financially able to do so. Hence, preferred stock tends to be issued when the firm is unable to borrow from other sources or the issuance of common stock creates ownership and control problems. For these reasons, relatively fewer firms issue preferred stock.

5.3 Common stock

Common stock may be sold privately to investors where flotation cost is less and public disclosure of information is not required. However, the sum raised may not be large enough, and institutional investors tend to impose stringent credit standards, voting control, and monitor the firm closely. For these reasons, common stock are often sold publicly in the primary market through an initial public offer (IPO), the organized secondary market (stock exchange) or the informal over-the-counter market for smaller firms (e.g. Nasdaq).

Common shareholders have last priority for payment, that is, after senior debt, operation and maintenance costs, and subordinated (or junior) debt holders are paid. Unlike debt, the dividend paid to common shareholders is not tax-deductible. In effect, common shareholders are taxed twice, once on corporate profit, and again on the dividends received. In return for the low priority in payment (and hence higher risk) and absence of tax benefits, equity investors may profit from dividends and capital gains.

Common shareholders have voting rights to elect the board of directors. They also have the right to receive assets upon dissolution of the firm. In a rights issue, shareholders are entitled to purchase additional shares (often at a discount or with "sweeteners" to encourage subscription) in proportion to their existing share holdings. For instance, in a 1-for-4 rights issue, a shareholder who currently owns 4,000 shares is entitled to subscribe to 1,000 new shares.

The maximum number of shares a firm can issue according to its charter is the authorized shares. The number of shares actually issued consists of outstanding shares held by the public and shares held or bought back by the firm (called treasury stock). The par value of a share is the stated amount of value per share in the charter.

Gordon's formula (see Chapter 2) may be used to determine the cost of common stock, that is,

$$r_E = g + (d_1/V).\tag{5.2}$$

where r_E is the rate of return to common stock (equity), d_1 is the first year dividend, and g is the annual dividend growth rate. The formula may also be written as

$$r_E = r_F + [g - r_F + (d_1/V)] = r_F + \lambda.\tag{5.3}$$

The advantage of rewriting it as Equation (5.3) is that it allows us to conclude that the risk premium in Gordon's formula depends on the excess dividend growth rate $(g - r_F)$ and first year return (d_1/V).

Example

If a firm's share is currently traded at $2 per share and the dividend of $0.06 is expected to grow at 3 per cent annually, what is the cost of common stock?

The solution is

$$r_E = 0.03 + 0.06/2 = 0.06 \text{ or } 6\%.$$

Instead of using Gordon's formula, the cost of equity may also be estimated using the more cumbersome Capital Asset Pricing Model (CAPM). The model is based on a mean-variance framework, that is, the return on an investment (r) is characterized by its mean and variance. The mean or average return on an asset is given by

$$E[r] = \mu\tag{5.4}$$

where $E[.]$ is the expectations or mean operator. The variance of the return is

$$\text{Var}(r) = \sigma^2.\tag{5.5}$$

The standard deviation (σ) is a measure of the dispersion of the returns and is therefore a measure of risk. Thus, if an investment yields a mean return of 6 per cent with a standard deviation of 2 per cent, one can write the return as $6 \pm 2\%$.

The mean-variance framework is only an approximation. It ignores skewness (degree of asymmetry) and kurtosis (degree of peakness) in the distribution of returns. Of the three distribution curves in Figure 5.2, *A* is skewed, *B* is relatively flat, and only C is the type of distribution assumed by the CAPM model. The returns from an investment need not be symmetrically distributed.

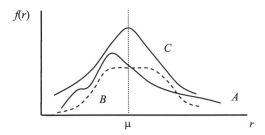

Figure 5.2 Distributions of returns.

Consider a portfolio comprising

- a risk-free asset (such as a T-bill) with return r_F and zero variance since it is risk-free; and
- a risky asset with return r, mean return $E[r]$, and $\text{Var}(r) = \sigma^2$.

The *portfolio return* is the weighted average of the two returns, that is,

$$R_P = (1 - w)r_F + wr \qquad\qquad (5.6)$$

where w is the portion of funds invested in the risky asset so that $(1 - w)$ is invested in the risk-free asset. The *expected portfolio return* is

$$E[R_P] = (1 - w)r_F + wE[r]. \qquad\qquad (5.7)$$

The expected portfolio return is basically the average return on the portfolio. Since the risk-free asset has zero variance (or standard deviation), its expected return is also r_F.

Example

If 30 per cent of my funds are invested in T-bills with a mean return of 4 per cent and the rest are invested in property with a 12 per cent average return, what is my expected portfolio return?

The solution is

$$\begin{aligned} E[R_P] &= (1 - w)r_F + wE[r] \\ &= 0.3(4\%) + 0.7(12\%) = 9.6\%. \end{aligned}$$

From Equation (5.6), the *portfolio variance* is

$$\text{Var}(R_P) = \text{Var}[(1-w)r_F + wr]$$
$$= (1-w)^2\text{Var}(r_F) + w^2\text{Var}(r) + 2(1-w)w\text{Cov}(r_F, r)$$
$$= 0 + w^2\text{Var}(r) + 0$$

i.e. $\sigma_P^2 = w^2\sigma^2.$ (5.8)

The first line in Equation (5.8) applies the variance operator Var(.) to both sides of
Equation (5.6). The second line uses the following result: if x and y are variables,
and c, d are constants, then

$$\text{Var}(cx + dy) = c^2\text{Var}(x) + d^2\text{Var}(y) + 2cd\text{Cov}(x, y) \qquad (5.9)$$

where Cov(x, y) is the covariance between x and y. That is,

$$\text{Var}(x) = \frac{1}{n}\sum(x_i - \mu_x)^2;$$

$$\text{Var}(y) = \frac{1}{n}\sum(y_i - \mu_y)^2; \text{ and}$$

$$\text{Cov}(x, y) = \frac{1}{n}\sum(x_i - \mu_x)(y_i - \mu_y). \qquad (5.10)$$

The summation runs from $i = 1$ to n, the total number of data points, and μ is the
population mean. If a small sample is used ($n < 30$), the denominator n should be
replaced by a smaller divisor ($n - 1$) because a small sample tends to under-
estimate the true (population) variance or covariance. Technically, we say that the
estimator with $n - 1$ as the divisor is unbiased, that is, its expected value is close to
the population parameter. For large samples, the use of n or $n - 1$ as the divisor
makes little difference. Irrespective of sample size, the population means are
replaced by sample means in computing sample variances and covariances.

Proof

Using Equation (5.10) and the definition of variance,

$$\text{Var}(cx + dy) = \frac{1}{n}\sum[(cx + dy) - (c\mu_x + d\mu_y)]^2$$

$$= \frac{1}{n}\sum[(cx - c\mu_x) + (dy - d\mu_y)]^2$$

$$= \frac{1}{n}\sum[c^2(x - \mu_x)^2 + d^2(y - \mu_y) + 2cd(x - \mu_x)(y - \mu_y)]$$

$$= c^2\text{Var}(x) + d^2\text{Var}(y) + 2cd\text{Cov}(x, y).$$

The third line in Equation (5.8) follows from the assumption that, for a risk-free asset,

$$\text{Var}(r_F) = 0; \text{ and}$$
$$\text{Cov}(r_F, r) = 0.$$

Thus, from the last line in Equation (5.8),

$$w = \sigma_P / \sigma. \tag{5.11}$$

Substituting w into Equation (5.7) gives, after some rearranging,

$$E[r] = r_F + \frac{\sigma}{\sigma_P}(E[R_P] - r_F). \tag{5.12}$$

If we let $R_P = r_m$, the market rate of return based on the stock market index, then the expected rate of return to the firm's equity investors is given by the following so-called security market line:

$$E[r_E] = r_F + \frac{\sigma}{\sigma_m}(E[r_m] - r_F) = r_F + \beta(E[r_m] - r_F). \tag{5.13}$$

Here $\beta = \sigma / \sigma_m$ is the firm's beta, the ratio of the standard deviation of the firm's return to the standard deviation of the stock market return. Thus, the risk premium in CAPM is

$$\lambda = \beta(E[r_m] - r_F).$$

From Equation (5.3), the risk premium in Gordon's model is

$$\lambda = g - r_F + (d_1/V).$$

The risk premiums differ.

In practice, Equation (5.13) is estimated by rewriting it as a simple linear regression model

$$E[r_E] - r_F = \alpha + \beta(E[r_m] - r_F) + \varepsilon. \tag{5.14}$$

That is,

$$y = \alpha + \beta x + \varepsilon \tag{5.15}$$

where

$$y = E[r_E] - r_F = \text{excess return on the firm's stock (i.e. return over and above the risk-free rate);}$$

$x = E[r_m] - r_F =$ excess market return; and

$\varepsilon =$ random error term.

The risk-free rate r_F may be approximated using the return on short-term T-bills rather than the long-term rate on government bonds. This assumes equity investors have short-term investment horizons, and long-term rates are not risk-free. The expected return on the firm's equity is computed using

$$E[r_E] = d + (\Delta P/P_{t-1})$$

where d is the dividend, and

$$\Delta P = P_t - P_{t-1}$$

where P_t is the firm's share price at time t. Often, $E[r_E]$ is computed using monthly data. Hence, the annual or semi-annual dividend needs to be adjusted accordingly. Finally, the expected stock market return $E[r_m]$ is computed using the stock market index (I), i.e.

$$E[r_m] = \Delta I/I_{t-1}.$$

As before,

$$\Delta I = I_t - I_{t-1}.$$

If monthly data are used, the number of data points should be at least 30 to obtain a reasonable estimate of β.

For each data point, the regression model is given by

$$y_i = \alpha + \beta x_i + \varepsilon_i \qquad i = 1,\dots, n \qquad (5.16)$$

The population parameters (α and β) and error terms ε are not observable. The sample estimates are a, b and residuals e respectively, that is, the estimated equation is

$$y_i = a + bx_i + e_i \qquad i = 1,\dots, n$$

Obviously, different samples give different estimates of a and b. This is why it is important to use an unbiased sample. A large sample should be used to reduce the bias.

As an illustration, suppose the monthly data are as follows:

Month	y(%)	x(%)
Jan	2	0
Feb	3	2
Mar	6	7
Apr	4	5

Note only $n = 4$ monthly data points are used to illustrate the computations. Substituting the data into Equation (5.16) gives

$$\begin{bmatrix} 2 \\ 3 \\ 6 \\ 4 \end{bmatrix} = \begin{bmatrix} 1 & 0 \\ 1 & 2 \\ 1 & 7 \\ 1 & 5 \end{bmatrix} \begin{bmatrix} a \\ b \end{bmatrix} + \begin{bmatrix} e_1 \\ e_2 \\ e_3 \\ e_4 \end{bmatrix} \qquad (5.17)$$

This may be written in matrix form as

$$\mathbf{y} = \mathbf{Xb} + \mathbf{e} \qquad (5.18)$$

where $\mathbf{y}$ is a 4 x 1 vector of observations, $\mathbf{X}$ is a 4 x 2 design matrix, $\mathbf{b}$ is a 2 x 1 vector of estimated coefficients, and $\mathbf{e}$ is a 4 x 1 vector of residuals. Notice $\mathbf{X}$ may be written as $[\mathbf{1} \quad \mathbf{x}]$ where $\mathbf{1}$ is a vector of ones and $\mathbf{x}$ is the second column of $\mathbf{X}$ in Equation (5.17). Then Equation (5.18) becomes

$$\mathbf{y} = a\mathbf{1} + b\mathbf{x} + \mathbf{e} = \mathbf{y}^* + \mathbf{e} \qquad (5.19)$$

where $\mathbf{y}^*$ is a linear combination (addition) of the vectors $\mathbf{1}$ and $\mathbf{x}$ (Figure 5.3). The columns of $\mathbf{X}$ span a vector space (shown as a plane), and $\mathbf{y}$ is not in the space. The least squares solution minimizes the residual vector $\mathbf{e}$, and this occurs if $\mathbf{e}$ is orthogonal (perpendicular) to the plane.

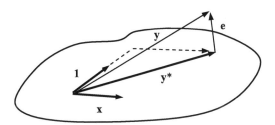

Figure 5.3 The geometry of least squares.

Premultiplying Equation (5.18) by the transpose matrix of $\mathbf{X}$ gives

$$\mathbf{X}^T\mathbf{y} = \mathbf{X}^T\mathbf{Xb} + \mathbf{X}^T\mathbf{e}. \qquad (5.20)$$

The transpose of a matrix is obtained by interchanging its rows and columns. That is,

$$\mathbf{X}^T = \begin{bmatrix} 1 & 1 & 1 & 1 \\ 0 & 2 & 7 & 5 \end{bmatrix}.$$

Now,

$$\mathbf{X}^T\mathbf{e} = [\mathbf{1}\ \mathbf{x}]^T\mathbf{e} = \mathbf{1}^T\mathbf{e} + \mathbf{x}^T\mathbf{e} = 0$$

since $\mathbf{e}$ is orthogonal to $\mathbf{1}$ and $\mathbf{x}$ (see Figure 5.3). Hence Equation (5.20) reduces to the *normal equations*

$$\mathbf{X}^T\mathbf{y} = \mathbf{X}^T\mathbf{X}\mathbf{b}. \tag{5.21}$$

The least squares solution is obtained by solving the normal equations, that is,

$$\mathbf{b} = (\mathbf{X}^T\mathbf{X})^{-1}\mathbf{X}^T\mathbf{y}. \tag{5.22}$$

For our monthly data,

$$\mathbf{X}^T\mathbf{X} = \begin{bmatrix} 1 & 1 & 1 & 1 \\ 0 & 2 & 7 & 5 \end{bmatrix} \begin{bmatrix} 1 & 0 \\ 1 & 2 \\ 1 & 7 \\ 1 & 5 \end{bmatrix} = \begin{bmatrix} 4 & 14 \\ 14 & 78 \end{bmatrix}; \text{ and}$$

$$\mathbf{X}^T\mathbf{y} = \begin{bmatrix} 1 & 1 & 1 & 1 \\ 0 & 2 & 7 & 5 \end{bmatrix} \begin{bmatrix} 2 \\ 3 \\ 6 \\ 4 \end{bmatrix} = \begin{bmatrix} 15 \\ 68 \end{bmatrix}.$$

Thus,

$$(\mathbf{X}^T\mathbf{X})^{-1} = \begin{bmatrix} 0.67 & -0.12 \\ -0.12 & 0.03 \end{bmatrix}$$

and, using Equation (5.22),

$$\mathbf{b} = (\mathbf{X}^T\mathbf{X})^{-1}\mathbf{X}^T\mathbf{y} = \begin{bmatrix} 0.67 & -0.12 \\ -0.12 & 0.03 \end{bmatrix} \begin{bmatrix} 15 \\ 68 \end{bmatrix} = \begin{bmatrix} 1.89 \\ 0.24 \end{bmatrix} = \begin{bmatrix} a \\ b \end{bmatrix}.$$

Hence, the estimated value of the firm's beta is 0.24. In practice, it is a simple matter to use statistical software (e.g. a spreadsheet) to run the above regression model. The hardest part is to get adequately long time series data on the firm's return and the market return based on a stock index.

How do we interpret it? Since $\beta = \sigma/\sigma_m$, it is equal to 1 for the stock market index. A stock is said to be "aggressive" if its price fluctuates more than the market index, that is, $\sigma > \sigma_m$ and such stocks have $\beta = \sigma/\sigma_m > 1$. A stock with $\beta < 1$ is "defensive."

Example

If a firm's beta is 0.24, the risk-free rate is 3 per cent, and the long-term expected market return is 7 per cent, what is the rate of return to equity for shareholders of the firm?

The solution is

$$E[r_E] = r_F + b(E[r_m] - r_F) = 3 + 0.24(7 - 3) = 3.96\%.$$

The low return merely reflects the fact that the firm's stock is defensive, that is, less risky.

We have seen that there are two ways in which the cost of common stock may be estimated:

Gordon's formula: $r_E = g + (d_1/V)$.
CAPM: $E[r_E] = r_F + \beta(E[r_m] - r_F)$.

Which one should we use? Gordon's formula is easier to use (and hence more popular) since it requires only estimates of the annual dividend growth rate (g), the current dividend (d_1), and the firm's current share price (V).

CAPM requires an estimate of the risk-free rate and time series data on the market index and the firm's share price from which rates of return are computed. It is then necessary to run a simple regression model using statistical software. The main weaknesses of CAPM are

- it is a single-factor model, and hence neglects factors such as price to earnings ratio (found in Gordon's model), country risk (say λ_C) or macroeconomic variables that may be added to the right hand side of the CAPM equation;
- the results are unreliable if the proxy stock market index used to compute market returns is inefficient so that prices do not fully reflect all the available information about the firm (Roll and Ross, 1994); and
- beta may be unstable over time, which is expected in volatile stock markets (Bos and Newbold, 1984).

Given these limitations, it is not surprising that empirical support for the CAPM model is weak (Davis, 1994; Fama and French, 1992) and it has limited use in estimating the cost of equity. Most analysts prefer to use the simpler Gordon's formula.

5.4 New issues

If a firm cannot borrow from other sources to pare down its debt or invest in new projects, it may issue new shares to existing shareholders (called rights issue) at a discount to market value to entice them to buy up the shares (i.e. exercise the rights).

Stock rights, which are options to purchase securities at a specific price at a future date, are issued in proportion to the number of shares held by an investor to maintain voting control and prevent dilution of ownership and earnings. These rights are tradable; existing shareholders may exercise them or sell it to other parties. From Gordon's formula, the cost of a new issue is

$$r_E = g + (d_1/V_D)$$

where V_D is the discounted share price to market value.

Example

A firm currently pays a dividend of $1 that is not expected to grow. If the current share price is $21 and new issues are sold at $20.10 with a flotation cost of $0.10 per share, what is the cost of the new issue?

Here
$$r_E = g + (d_1/V_D) = 0 + 1/20 = 5\%.$$

If the above rate of return is too low, the firm needs to raise its dividend or lower V_D to say $19.10. Then $r_E = 1/19 = 5.3\%$. The number of new shares to issue depends on

- the number of outstanding shares held by the public (e.g. 10,000,000);
- the issue price ($19.10); and
- the amount to be raised (e.g. $40 m).

Thus,

Number of new shares = ($40 m)/$19.10 = 2,094,241
Ratio = 10,000,000/2,094,241 = 4.8

The firm needs to issue 1 share for every 5 shares held.   Thus, if you own 5,000 shares and the firm decides to have a one-for-five rights issue, you are entitled to buy an additional 1,000 shares. Your holdings are as follows:

5,000 shares $21 per share	$105,000
1,000 new shares at $19.10 per share	$19,100
Value of 6,000 shares	$124,100
Ex-rights value per share	$124,100/6,000 = $20.68.

This means that the share price will (theoretically) fall to $20.68 after the rights issue.

5.5 Bonds

The cost of issuing a bond depends on the type of bond, maturity, general level of interest rates, and issuer (i.e. risk). A plain vanilla or fixed rate bond pays a coupon (interest) each period and, at maturity, the last coupon payment is made along with the par value (i.e. stated nominal value, or principal). Generally, investment grade bonds with longer maturities have higher coupon rates to compensate holders for being exposed to inflation and other risks for a longer period of time.

Suppose a firm issues a 10-year bond with 5 per cent coupon (i.e. stated annual interest) at a par value of $1,000. The bond is sold at a discount at $980, and floatation cost is $20. Thus the net proceeds is $980 − $20 = $960. Each year, the firm pays a coupon of $0.05 \times \$1,000 = \$50$ for each bond. At maturity, the firm pays the last coupon of $50 and principal ($1,000) for a total of $1,050. The cost of the bond is the value of r in

$$V = \frac{C_1}{1+r} + \frac{C_2}{(1+r)^2} + \cdots + \frac{C_n}{(1+r)^n} \qquad (5.23)$$

i.e.

$$960 = \frac{50}{1+r} + \frac{50}{(1+r)^2} + \cdots + \frac{1050}{(1+r)^{10}}.$$

The equation may be solved using trial and error on a spreadsheet. Different trial values of r are used until the right hand side (RHS) is close to 960. Some trial numbers are given below:

Trial value of r_B	RHS
0.10	692
0.05	1,000
0.06	926
0.055	962

When the first trial value of 0.10 is substituted for r on the RHS, the value of the sum is 692, which is well below 960. This means that 0.10 is too high, and when 0.05 is used, the value of the sum rises to 1,000, indicating that r is between 0.05 and 0.10, but closer to 0.05. The next trial value gives a RHS value of 926, suggesting that r is between 0.05 and 0.06. The final estimated value of r is 0.055 or 5.5%. As this example shows, the iteration should converge after a few trials because the function is well-behaved. More complex numerical methods such as Newton's method may be used to compute r. However, Newton's method requires calculus and is cumbersome to use if there are many terms on the right hand side of the equation. Hence, for all practical purposes, the trial and error method works just fine.

Debt is cheaper than equity for three reasons, namely,

- interest on debt (bonds and loans) is tax deductible for the purpose of computing corporate tax to encourage investment. If t is the corporate tax rate, the effective cost of bond is

$$r_B = (1 - t)r;$$

- priority over equity in distribution of earnings so that lenders require a lower rate of return; and
- lower transaction cost.

Of the three reasons, the most important for the widespread use of corporate debt is tax deductibility. If the corporate tax rate is 30 per cent, the effective cost of the bond is only $0.7(5.5\%) = 3.85\%$.

If interest is not paid annually, the valuation equation must be adjusted as shown in the example below.

Example

Find the value of a bond with 3 years remaining if the 6 per cent coupon is paid semi-annually and its par value is $1,000. What happens to the value of the bond if interest rate rises to 8 per cent?

The semi-annual interest is $0.06(1,000)/2 = \$30$. The semi-annual interest rate is $6/2 = 3\%$, and the number of compounding periods is $3 \times 2 = 6$. Hence,

$$V = 30/1.03 + 30/1.03^2 + \cdots + 1030/1.03^6$$
$$= \$1,000.$$

If the interest rate rises to 8 per cent, the semi-annual interest rate is $8/2 = 4\%$, and value of the bond is

$$V = 30/1.04 + 30/1.04^2 + \cdots + 1030/1.04^6$$
$$= \$948.$$

As expected, the value of a bond falls when interest rate rises because of the inverse relation between V and r in Equation (5.23).

Corporate bonds may or may not be subordinated to senior debt. Subordination means that the senior debt will have priority in debt servicing relative to the subordinated or junior debt. Hence, if a bond is subordinated to a bank loan, then the corporation will have to service the bank loan first through periodic repayments before serving the bond through dividends. As discussed in Section 5.3, all debts, irrespective of whether they are loans or bonds, are serviced only if the firm is financially able to do so. That is, revenue must first be used to pay for operations and maintenance before debt service. Equity investors who are paid dividends have last priority for payment.

Apart from the plain vanilla bond, there are many other types of bonds. Debentures are generally unsecured corporate bonds. This makes it attractive for a firm to issue such bonds but at higher cost to compensate bond holders for the higher risk. Sometimes, bonds and debentures are used interchangeably, so care is required in interpreting the specific terms of a bond issue. Some debentures are secured against the asset.

Mortgage bonds and collateral trust bonds are secured by the real and financial assets of the issuer respectively. Airlines and railway companies may also issue equipment trust certificates ("bonds") secured by equipment. Other types of bonds include

- zero coupon bonds that pay no periodic coupon (interest) but are sold at a steeper discount;
- floating rate bonds with variable coupon rates based on a benchmark interest rate such as the London Interbank Offered Rate (LIBOR) or Singapore Interbank Offered Rate (SIBOR);
- inverse floater bonds that pay a coupon rate determined by a fixed rate (e.g. 7 per cent) less LIBOR (or SIBOR);
- low or speculative grade junk bonds with higher yields that are issued by a company with higher credit risks, possibly to finance corporate mergers and takeovers;
- tax-exempt municipal bonds issued by local or state governments to raise money to finance infrastructure;
- short-term extendible bonds that are extendible on maturity; and
- putable bonds that are redeemable at par value at the option of the holder at specific dates or under certain conditions.

Bonds may be issued nationally, regionally or globally depending on the size of the issue. Examples of foreign bonds include Yankee, Bulldog, Samurai or Global bonds. If it is denominated in a currency different from the currency of the country of issue, it is a called Eurobond. It has nothing to do with the Euro currency or the European bond market. Bonds are usually callable, that is, the issuer may repurchase it prior to maturity at the stated call price. This may happen if a new bond could be reissued at a lower coupon rate.

5.6 Bank loans

Senior debt consists of loans provided by commercial banks. Banks also provide short-term loans, credit lines, and letters of credit used in international trade to pay a seller.

The interest rate charged on a bank loan varies according to the demand and supply of loanable funds, credit worthiness of the borrower, viability of the project, and maturity.

Both fixed rate and variable rate interest may be charged depending on who bears the inflation risk. If a fixed rate is used, the lender bears the risk and hence fixed rate loans tend to command higher interest rates than variable rate loans where inflation risk is shifted to the borrower. In the past, lenders tended to offer long-term fixed rate loans. However, the high inflation of the 1970s changed all that. As inflation rose, so did deposit rates offered to savers and banks that "borrowed short and lent long" (particularly on mortgage loans) lost money because the short-term rates offered to savers were higher than long-term fixed rates charged to borrowers. Nowadays, many lenders prefer variable rate loans. This shifts inflation risk to the borrower who, paradoxically, may not be the best person to hedge such risks.

Banks tend to charge higher interest rates for loans with longer maturities because of greater risk exposure. However, they may use short-term credit to control borrowers (Diamond, 1991) or if it is more difficult to enforce loan covenants (Hart and Moore, 1995). Thus, short-term loans provide banks the flexibility to terminate or restructure projects that are not viable. This practice also provides corporate management with the incentive to avoid bad projects and respond more rapidly to adverse shocks (Ofek, 1993). It is sometimes alleged that investors (i.e. lenders or "Wall Street") tend to take a short term view of corporate investment for paper gains and this prevents companies from investing in the long term to develop their core competences. It is too easy to blame Wall Street for a company's woes.

5.7 Pseudo-equity

Junior debt as layered or mezzanine finance is often treated as pseudo-equity. Such debt may be unsecured, that is, they may not be backed by any collateral. An example is the issuance of short-term corporate commercial paper to meet short-term liabilities.

Generally, project revenues are used to pay for the following items in order of priority:

- operations and maintenance;
- senior debt service;
- senior debt service reserve;
- junior debt service; and
- other project reserves.

Since junior debt has lower priority in debt servicing, it is more risky. Hence, it has a higher yield than senior debt. Note the use of debt service and project reserves to alleviate possible cash flow problems.

5.8 Weighted average cost of capital

Since the firm borrows from different sources of funds at different rates, its weighted average cost of capital (WACC) is

$$\text{WACC} = \sum w_i r_i \qquad\qquad (5.24)$$

where w_i is the proportion of funds borrowed from the ith source and r_i is the cost of each source of capital. A simple example will make this clear.

Example

Suppose a firm borrows the following amounts from different sources at various effective costs (i.e. adjusted for tax deduction on debt interest). Compute the WACC.

Source	Proportion (w_i)	Effective cost of funds (r_i)
Bank loans	0.4	5%
Bonds	0.3	5%
Preferred stock	0.1	6%
Retained earnings	0.2	7%
New issues	0.0	–
Total	1.0	

$$\text{WACC} = 0.4(5) + 0.3(5) + 0.1(6) + 0.2(7)$$
$$= 2.0 + 1.5 + 0.6 + 1.4 = 5.5\%.$$

5.9 Optimal capital structure

The WACC raises the question on the optimal structure of debt and equity for a firm. That is, is there an optimal mix of debt and equity that gives the lowest cost of capital for the firm?

If debt is cheaper than equity because interest on debt is tax deductible, then a firm should raise more debt to reduce its WACC. This is the traditional view of capital structure. However, using more debt increases the risk of default. Hence, if WACC is plotted against percentage of debt (leverage), there is a point where WACC is minimized (Figure 5.4).

In contrast to the traditional view of a U-shaped capital structure, Modigliani and Miller (1958) claimed that capital structure does not matter in a world with perfect information and no taxes. If a firm is highly levered, investors can offset this by adjusting their investment portfolios. Hence, capital structure does not matter, and the WACC curve in Figure 5.4 is not U-shaped but horizontal.

Critics argue that, in the real world, there are taxes, institutional constraints on borrowing, imperfect information, problems of control, and agency costs. In short, capital structure matters. A firm cannot raise 100 per cent debt without increasing the probability of default. Hence, the cost of capital rises as the firm is increasing levered. Further, lenders require equity investment to signal commitment, and different countries have varied ways in which lenders deal with borrowers and impose different loan covenants. In some countries, banks and firms are distinct market players and deal at arm's length. In contrast, German and Japanese banks tend to develop closer relations with the firms.

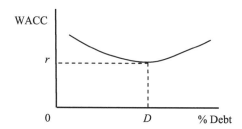

Figure 5.4 WACC and leverage.

If information is imperfect (asymmetric), the entrepreneur knows the project risk better than external investors and may be willing to pay more for external funds. In the extreme, this may encourage promoters to siphon money out of the project and then file for bankruptcy. Conversely, investors may perceive the project risk to be higher than the expectations of managers and therefore raise the cost of external debt.

There are also agency costs and control problems associated with raising debt and equity (Jensen and Meckling, 1976). Family-controlled firms may be reluctant to issue equity to avoid diluting control. If equity is issued in non-family controlled firms, ownership is also diluted, resulting in possible misalignment of managerial incentives. This misalignment raises the monitoring costs for external investors. There are also agency costs associated with the issue of debt. Since liability is limited, there may be an incentive for sponsors and management to use more debt (particularly if it is off the balance sheet) to undertake short-term and risky projects and walk away if a project fails.

Another issue in the optimal capital structure debate concerns priority of payment if a firm is close to bankruptcy. Since debt holders are paid first, shareholders may underinvest in projects if they do not reap sufficient benefit (Myers, 1977).

Finally, capital structure is also affected by the timing of cash flows. Firms need to time their expenditures and borrowings or equity issues and balance between short-term and long-term debt to match their assets and liabilities (Titman and Wessels, 1988).

In summary, capital structure matters because

- debt is cheaper than equity due to interest deductibility for tax purposes;
- lenders require equity investment from borrowers to signal commitment;
- banks and firms have differing relations across countries;
- information is asymmetric and this encourages opportunistic behavior;

- there are agency costs and problems of control;
- debt holders and equity investors have different priority of payment if the firm is bankrupt and they may behave differently if a firm is close to bankruptcy; and
- firms need to use a combination of debt and equity to manage their cash flows.

With these considerations, it is not surprising that capital structures differ substantially across firms and countries (Rajan and Zingales, 1995; Booth et al., 2001).

Questions

1 What are the major sources of funds for a firm?

2 Explain why preferred stock is seldom issued by firms.

3 Explain why firms may prefer to issue debt over equity, and vice versa.

4 A firm's share is currently trading at $10 per share and it pays a dividend of $0.50 per share with a growth rate of about 1 per cent each year. What is its cost of equity? [6%]

5 Instead of using Gordon's formula is Question 1, the firm tries to estimate its cost of equity using CAPM. Historically, its share price tends to fluctuate at roughly the same volatility with the market index, that is, the standard deviations are similar. If the long-term annual market return is 7 per cent and the risk-free interest rate is 3 per cent, what is the firm's cost of equity? [7%]

6 What are the strengths and weaknesses of CAPM?

7 It is often argued that CAPM is not useful because stock markets are more volatile in globalized and deregulated financial markets. Discuss.

8 A firm has an outstanding 10-year bond with 2 years to expiry. It has a par value of $1,000 and an 8 per cent coupon payable semi-annually. Compute the value of this bond. [$1,000]

9 What should a firm consider in issuing bonds to raise funds?

10 A firm currently uses 30 per cent equity and 70 per cent debt to finance its operations. If the cost of equity is 6 per cent and the cost of debt is 8 per cent, what is its WACC? [7.4%]

11 Why does capital structure matter?

12 Explain why it is difficult to determine a firm's optimal capital structure.

13　Common shareholders elect a Board of Directors to represent their interests in a firm. Explain why these shareholders may still not be able to fully monitor management through its Board.

14　In October 2006, property developer CapitaLand issued a $430 m 10-year bond with a 2.1 per cent coupon rate a year. The bond may be converted to shares at $7.31 per share at maturity, higher than its current share price of $5.25. Why is there demand for a bond issue with such a low annual coupon and high conversion premium at maturity?

6

Project Development

6.1 Owner's need

This chapter provides a brief overview of project development for readers who are not sufficiently familiar with the project life cycle to appreciate the role of finance in projects.

The term "owner," "client," or "sponsor" rather than the legalistic "employer" or "customer" will be used interchangeably in this chapter as well as this book unless otherwise stated.

Organizations invest in projects for

- strategic reasons such as to penetrate a market, take on new competitors, or introduce a new product; and
- tactical (routine) reasons.

Individual projects may stand alone or are embedded in a program of related projects. In the latter, projects are dependent or complementary.

A project begins with an identified owner's (or client's) need. It may be an upgrading or expansion of an existing facility or development of a new facility to capitalize on market opportunities.

It is assumed that the project is politically, technically, and financially feasible. Some projects such as a new dam may be politically sensitive because of possible destruction of wildlife, forest, and flooding. Road projects may also draw a similar response from compulsory acquisition of private land, inadequate compensation, delays, and destruction of nature.

In such cases, it is important to identify stakeholders such as local, regional and federal State agencies, management, lenders, insurers, the project team, contractors, subcontractors, suppliers, consumers, workers, affected residents, businesses, professionals, and the mass media. Stakeholder management is discussed in Chapter 8.

A project's technical feasibility will have to be worked out either by the owner or bidders. In general, owners are risk averse and do not like to take chances with untested technology.

The financial feasibility of a project is discussed in some depth later in this chapter.

6.2 Request for Proposal

The owner's need is often translated into a Request for Proposal (RFP) to potential bidders. Generally, the RFP contains the following:

- Covering letter;
- General description of the facility or items to procure;
- Requirements and specifications;
- Budget;
- Procurement method;
- Cost breakdown;
- Time schedule;
- General and administrative clauses; and
- Data and drawings (if appropriate).

The covering letter is brief and provides some basic information about the organization and project drawn from other parts of the RFP. The "General and administrative clauses" include items such as equal opportunity employment and use of part-time staff. Similarly, data and drawings will be provided to guide bidders. The rest of the items are discussed below.

Usually, a Request for Qualifications (RFQ) is issued prior to an RFP to shortlist suitable bidders.

6.3 General description of facility

The "General description of the facility" section forms a key part of an RFP. The scope of the project in terms of goals, objectives, and deliverables are spelt out in this section.

In many cases, an organization is able to define the scope for similar projects such as when a developer decides to build a new office block. The goal is clearly profitability, and the objectives are what the developer sets out to achieve, such as to enter the office market or make a presence in the Central Business District through a signature building.

However, a university may not have sufficient experience in building a new college. The project scope will then need to be carefully defined from a survey among stakeholders such as university senior management, teachers, administrative staff, and students.

An ill-defined project scope is a major source of scope creep, design changes, variation orders, and disputes. The scope may be ill-defined because the owner or consultant is inexperienced, or the owner is still wrestling with the concept for the facility.

Example

As an example, the general description of a residential building may include the items listed below.

1 General

Background of owner (firm), objectives of the project, status of site (e.g. awaiting completion of sale), likely uses of site, lot size, layout, possible design concepts, design issues, and provision for future expansion.

2 Site information

Location, land tenure, easements, covenants, zoning, utilities, access to transport and other facilities, climate, geology, soil condition, possible contamination, existing structures, topography, vegetation, and externalities such as air and noise pollution.

3 Building

3.1 Architectural

- Number of buildings, types, and sizes
- Breakdown of space uses
- Special features

3.2 Mechanical and electrical

- Power supply
- Plumbing, sanitation, drainage, and gas
- Lighting
- Heating, ventilation, and air-conditioning
- Fire protection
- Telecommunications
- Security systems
- Vertical transport

3.3 Landscaping

- Softscape – trees, plants, and ground cover
- Hardscape – walkways, pools, and so on

4 Special features

- Lighting
- Acoustics
- Interior design
- Information technology (e.g. wireless connectivity)
- Equipment

6.4 Requirements and specifications

There are various types of requirements for a facility or equipment, namely,

- program requirements that outline the functional needs to be fulfilled by the project and are often expressed in terms of quantitative data (e.g. number of blocks, number of units, and so on);
- performance requirements are more detailed and are described in the specifications and may include design standards, quality of finishes, quality of work, size requirements (e.g. parking), performance criteria, and methods of construction;
- statutory requirements including planning approvals, construction permits, environmental regulations, building regulations (structural, health, fire, and safety), and conservation restrictions; and
- Technical and Design Proposal Requirements (TDPR) that identify specific design and technical proposals bidders must provide (e.g. geotechnical issues). This requirement is necessary to avoid wasting time in evaluating unwanted technical proposals from bidders.

A simple example of these two types of requirements is the need for a computer as a functional requirement, and the specifications may include the brand, color, type of screen, speed, and size of memory.

The program requirements may be listed in the "General description of the facility" section.

6.5 Budget

The budget for a facility typically includes the following items:

- Land cost
- Land transaction cost (stamp duty and legal fees)
- Pre-construction cost
- Statutory fees (written permission and building plan approval)
- Soil investigation
- Site survey
- Professional fees (paid progressively from pre-construction stage)
  (Indicative fees as a percentage of construction cost are given in brackets)

  Architect (3%)
  Quantity surveyor (0.5%)
  Project manager (0.5%)
  Landscape architect (0.5%)
  Interior designer (0.5%)
  Mechanical & Electrical consultant (1%)
  Civil and structural consultant (1%)
  Others (0.5%)

- Construction cost based on preliminary estimates (e.g. $/m^2$) before detailed design is completed
- Marketing fees
- Legal fees
- Post-completion expenses
- Client's overhead
- Interest on land and construction loans
- Contingency
- Goods and services tax or value-added tax

In estimating the budget, the client will review vendor products and pricing through Requests for Information (RFI). Since suppliers who provide such information incur some costs, it is important to write an RFI clearly as suppliers may sense a lack of commitment on the owner's part and view the RFI as a "fishing expedition."

6.6 Financial feasibility

Whether a project is financially feasible depends on the criteria used. If a project incurs an initial cost C_0, the net present value is given by

$$\text{NPV} = -C_0 + \frac{N_1}{1+r} + \frac{N_2}{(1+r)^2} + \cdots + \frac{N_n}{(1+r)^n} \tag{6.1}$$

where N_t, $t = 1,..., n$ is net operating income (NOI) at the end of year t, and n is the terminal year, and r is the discount rate.

A project is worth considering if its NPV exceeds zero. Recall that NOI is obtained by deducting from sales revenue the cost of goods (e.g. direct labor, energy, and material costs), operating expenses (such as insurance, advertising, research and development, utilities, transport, and rent), and depreciation of buildings, vehicles, and equipment. A detailed example of these computations is given later in the section.

The initial cost may not be paid at the start but is disbursed over the first few years of construction. Hence, N_1, N_2 and N_3 may be negative as the facility is still under construction and earns no revenue. Further, if the facility has a salvage value at the end of n years, it should be included in N_n.

The salvage value is either estimated at future market value (inclusive of depreciation) or agreed beforehand between the parties. For instance, in a build-operate-transfer project, there may be an initial agreement for the government to pay the project sponsors (i.e. current owners) a certain sum of money for the facility when it is transferred to the State at the end of the concessionary period (e.g. 20 years).

Example

If a project has an initial cost of $30 m and can generate a net operating income of $10 m a year over 4 years with no salvage value at the end of 4 years, what is its NPV, assuming a discount rate of 5 per cent?

Using Equation (6.1),

$$\text{NPV} = -30 + 10/(1.05) + 10/(1.05)^2 + 10/(1.05)^3 + 10/(1.05)^4$$
$$= \$5.46 \text{ m} > 0.$$

The project is worth considering.

There are several weaknesses with the NPV criterion. First, NPV is merely an absolute number (e.g. $5.46 m) and hence inferior to rates of return (e.g. 10 per cent) because it does not scale for project size. For example, an NPV of $1 m for a $10 m project should be viewed differently from that of a $100 m project. Second, the discount rate is required to compute the NPV. It is not easy to determine the appropriate discount rate because it is difficult to ascertain the risk premium for the project. Third, it is assumed that forecasts of net operating incomes and salvage value are reasonably accurate. This assumption is harder to defend if markets are volatile or if distant projections are used or required.

For these reasons, the following internal rates of return (IRR) are more popular:

- project IRR; and
- equity IRR.

Unlike NPVs, IRRs give a rate of return rather than an absolute dollar value. They also do away with the need to specify an appropriate discount rate. However, IRRs need to be compared with a hurdle rate (discussed later in this section) above which a project is deemed to be feasibility. Finally, the problem of projecting distant net operating incomes and salvage value remains when IRR is used.

Project IRR

The project IRR is also called the "free and clear" IRR because it assumes the project is unlevered, that is, debt financing is ignored or the project is "free and clear" from debt. It is the discount rate k that makes NPV equal to zero. Hence we solve

$$0 = -C_0 + \frac{N_1}{1+k} + \frac{N_2}{(1+k)^2} + \cdots + \frac{N_n}{(1+k)^n} . \tag{6.2}$$

Example

If a project has an initial cost of $30 m and can generate a net operating income of $10 m a year over 4 years with a $2 m salvage value at the end of 4 years, what is its project IRR?

Using Equation (6.2),

$$0 = -30 + \frac{10}{1+k} + \frac{10}{(1+k)^2} + \frac{10}{(1+k)^3} + \frac{12}{(1+k)^4}$$

This equation may be solved by using trial values of k until the right hand side (RHS) is close to the left hand side (i.e. zero). Usually, a spreadsheet is used to avoid careless mistakes.

Trial value of k	RHS
0.10	3.06
0.15	−0.31
0.145	0.0046
0.1451	−0.0017

The first trial of $k = 0.10$ gives a RHS value of $3.06 m, which is above zero. The second trial of $k = 0.15$ gives a RHS value of −$0.3 m, which is below zero. Hence, k lies between 10 and 15 per cent. The estimated value of k is 14.51 per cent.

We note in passing that there are two other methods of finding the IRR that are sometimes used but are not recommended here. One method is to interpolate between two trial values of k such as k_1 and k_2 in Figure 6.1. Note that k_1 must result in a positive NPV and k_2 must give a negative NPV for the interpolation to make sense. The interpolation method is generally less accurate than the trial and error method described above.

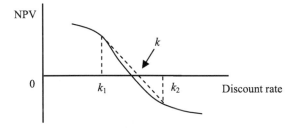

Figure 6.1 Interpolation method.

The other approach to find the root of the equation (i.e. IRR) is to use Newton's method. It requires the use of calculus and is therefore cumbersome to use if there are many terms on the right hand side that need to be differentiated to find the gradient. Further, the process may not converge if the initial guess value is far from the solution.

Newton's method is shown in Figure 6.2. Staring from an initial guess k_1, we compute the gradient and use it to find k_2. At k_2, we compute the slope again and use it to find k_3, and so on until convergence. For the process in Figure 6.2, the convergence is rapid. However, if the curve is relatively flat near the root, the gradient may be nearly horizontal and does not cut the x-axis at all. The process diverges. The main disadvantage with Newton's method in our context is that it is cumbersome to use.

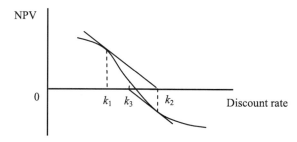

Figure 6.2 Newton's method.

Equity IRR

The use of project or "free and clear" IRR does not consider how a project is financed using debt and equity.

If debt is used, the appropriate rate of return is not the rate of return to total capital (i.e. project IRR) but equity IRR. The initial cost (C_0) and net operating incomes (N_t) in Equation (6.1) are replaced by initial equity (E_0) and cash flows (F_t) respectively. The project IRR (k) is replaced by the equity IRR (q) so that

$$0 = -E_0 + \frac{F_1}{1+q} + \frac{F_2}{(1+q)^2} + \cdots + \frac{F_n}{(1+q)^n} .$$ (6.3)

As before, trial values are used to solve for q.

Example

A facility costs $10 m to build and is financed using $2 m equity ($E_0$) and $8 m debt at 5 per cent interest for 15 years. If it is sold at the end of 4 years for $12 m, what is the equity IRR?

The cash flow for each year is found using Table 6.1. It is assumed that the facility is still under construction in the first year, and the initial equity of $2 m is paid before construction begins and the $8 m loan is disbursed during construction in Year 1. However, debt repayment begins at the end of the first year.

The annual depreciation (d) is computed using the straight line method (see Chapter 4), that is,

$$d = (\text{Original structure cost} - \text{scrap value})/\text{depreciation period}.$$

			Year		
	0	1	2	3	4
Initial cost	−2.000				
Sales revenue			3.000	3.000	3.000
Less: Cost of goods sold			1.000	1.000	1.000
Gross profit			2.000	2.000	2.000
Less: Operating expenses			1.000	1.000	1.000
Less: Depreciation			0.200	0.200	0.200
Net operating income (NOI)			0.800	0.800	0.800
Other income			0	0	0
Earnings before interest and tax			0.800	0.800	0.800
Less: Interest expense		0.400	0.381	0.362	0.342
Earnings before tax (EBT)			0.419	0.438	0.458
Less: Tax@20%			0.084	0.088	0.092
Net earnings			0.335	0.350	0.366
Add: Depreciation			0.200	0.200	0.200
Less: Principal payments		0.371	0.389	0.409	0.429
Cash flow	−2.000	−0.771	0.146	0.141	0.137

Table 6.1 Cash flow ($m).

Year (a)	Principal at start of period (b) ($)	Annual repayment (c) ($)	Payment breakdown	
			Interest (d) ($)	Principal (e) ($)
1	8,000,000	770,738	400,000	370,738
2	7,629,262	770,738	381,463	389,275
3	7,239,987	770,738	361,999	408,739
4	6,831,248	770,738	341,562	429,176
5	6,402,072	770,738	320,104	450,634
Etc				

Table 6.2 Loan amortization schedule.

Since land cannot be depreciated, the original structure cost is computed by subtracting an estimate of land value ($3 m say) from the total cost of facility (i.e. $10 m). The scrap value of the facility is assumed to be zero, and the depreciation period allowed by the tax authority for this facility is 35 years. Hence,

$$d = (7 - 0)/35 = \$0.2 \text{ m}.$$

For simplicity, it is assumed that no other assets (e.g. vehicles and equipment) are depreciated. If there are, the depreciation may be computed in a similar manner using the appropriate allowable depreciation period given by the tax office. The total depreciation amount is then the sum of depreciated amounts for different fixed assets.

Since interest expense is tax deductible for corporate tax purposes, it is deducted from net operating income (and other income) to determine earnings before tax (EBT). It is assumed that EBT is taxed at 20 per cent.

The next step is to compute net earnings by subtracting corporate tax from EBT. Since depreciation is actually not expensed, it is normally (or arguably) added back to net earnings. Finally, principal payments are then deducted to obtain the cash flow. Note that only the principal payments are deducted; the interest portion has been deducted earlier as interest expense.

The periodic interest expense and principal payments in Table 6.1 are computed from the amortization table in Table 6.2. Column (b) shows the principal of $8m at the start of Year 1. The *annual* loan repayment is computed using

$$z = \frac{Li}{1-(1+i)^{-n}} \tag{6.4}$$

where L is the principal, i is the loan interest rate ($=0.05$), and n is the term of the loan (15 years). We note in passing that the *monthly* repayment is not found by dividing z by 12 but by dividing the annual interest rate (i) by 12 and multiplying the number of repayment periods (n) by 12.

In our example, the annual loan repayment is

$$z = \frac{8(0.05)}{1-(1+0.05)^{-15}} = \$0.770738m = \$770,738.$$

The interest portion of the repayment in column (d) for each row is found by multiplying column (b) by the rate of interest. For example, for the first row,

$$8,000,000(0.05) = \$400,000.$$

The last column (e) is obtained using (c) – (d), i.e. annual repayment less interest. Hence, for the first row,

$$770,738 - 400,000 = \$370,738.$$

For the second row, the principal at start of period is found using (b) – (e) from the first row. Hence,

$$8,000,000 - 370,738 = \$7,629,262.$$

The other entries in the second row are computed in the same manner as described above. The interest portion of the loan in Table 6.2 for each year are then transferred to Table 6.1 under "interest expense" but are converted to $m and rounded to three decimal places.

At the end of 4 years, the facility is sold for $12 m (net of transaction cost) of which the remaining principal of $6.402 m (see column (b) of Table 6.2) needs to be repaid. Hence, the capital gain is $12 m − $6.402 m = $5.598 m, and it is assumed that this is not taxed.

Using Equation (6.3), we have

$$0 = -E_0 + \frac{F_1}{1+q} + \frac{F_2}{(1+q)^2} + \cdots + \frac{F_4}{(1+q)^4}$$
$$= -2 - \frac{0.771}{1+q} + \frac{0.146}{(1+q)^2} + \frac{0.141}{(1+q)^3} + \frac{5.598 + 0.137}{(1+q)^4}.$$

Using trial and error, $q = 24.5\%$.

Hurdle rate

Recall from Chapter 5 that a firm has a weighted average cost of capital (WACC). The WACC is an indicator of the evaluation by lenders and investors on the riskiness of the *entire* firm based on a mixture of debt and equity.

The project IRR should exceed WACC, and the equity IRR should exceed the firm's cost of equity. However, we saw from Chapter 5 that the rate of return to equity investors is given by the security market line (SML)

$$E[r_E] = r_F + \beta(E[r_m] - r_F) = r_F + \lambda \tag{6.5}$$

where λ is the risk premium. The SML is plotted in Figure 6.3. The expected market return $E[r_m]$ corresponds to where the SML cuts $\beta = 1.0$. This can be seen by substituting $\beta = 1.0$ into Equation (6.5) so that

$$E[r_E] = r_F + \beta(E[r_m] - r_F) = E[r_m].$$

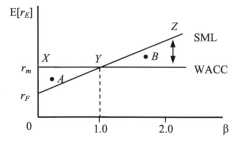

Figure 6.3 Security market line.

The expected returns to equity from two projects (*A* and *B*) are also plotted. If WACC is used as the financial investment criteria, then project *A* is not viable. However, the project risk is low, and the expected return is actually above that required by equity investors as shown by the SML.

In contrast, the expected return from project *B* is above WACC and is therefore viable under this criterion. However, it is below the expected return required by equity investors.

Hence, the use of WACC may not be consistent with the Capital Asset Pricing Model (CAPM). It does not require project managers any knowledge of CAPM to realize that a mark-up should be used so that

Project hurdle rate = WACC + mark-up.

The mark-up is required to compensate the firm for project-specific risk (e.g. country risk). It is shown as a double-headed arrow in Figure 6.3, and the intent is to approximate the SML. Since most firms would not approve of projects below WACC, the financial criterion is actually given by the kinked line *XYZ* where WACC is used for less risky projects (up to point *Y*) and an approximate SML is used for riskier projects (from *Y* to *Z*).

The long-term rather than short-term WACC is used as the floor. It does not make sense for the firm to keep adjusting its WACC just because its cost of capital happen to fluctuate each time it raises capital. This is the so-called "separation principle" where the hurdle rate used is independent of how a project is financed. However, we have also explored in Section 5.9 various reasons why capital structure matters.

6.7 Project authorization

Once a project is deemed to be politically, technically, and financially feasible, it is authorized by management.

The Project Charter consists of the following:

- Background;
- Current system or facility;
- Proposed system or facility;
- Scope;
- Budget;
- Timelines and phases;
- Project manager;
- Roles and responsibilities of key members; and
- Stakeholders.

Once a project has been authorized, the appointed project manager develops a preliminary organization structure and assembles the project team for a kick-off meeting to brief team members of the tasks ahead.

6.8 Procurement method

The procurement method is specified in the Request for Proposal. Some common methods are outlined below (see Brook, 2004).

Traditional method

In the traditional design-bid-build method of delivery (Figure 6.4), the project team comprising the project manager, design specialists (architects and engineers), cost engineer, and so on. The team is appointed by the owner. If the owner is a large development firm, some members of the project team (e.g. project manager) may be employed directly (i.e. in-house).

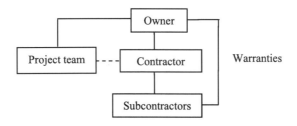

Figure 6.4 Traditional delivery method.

The owner engages the project team to design the project. After the design has been completed, pre-selected contractors are invited to tender for the project based on detailed design drawings, bills of quantities, specifications, and a selected form of contract. The award of contract is based on

- lowest bid, or
- a pre-determined set of criteria (price, design, track record, and so on).

Subcontractors may be nominated by the owner (called nominated subcontractors) or appointed by the contractor (called domestic subcontractors). Apart from the warranties given to the owner, there is no contractual relation between the owner and subcontractors.

The advantages of this system are:

- the project cost is more or less fixed because the design has been finalized and known to the owner before tendering and construction; and
- the project team does not have a contractual relation with the contractor (shown dotted in Figure 6.1). The team supervises the construction on behalf of the owner.

The disadvantages are:

- there is no design input from the contractor prior to tender;
- the separation of design and construction; if anything goes wrong, the design team may blame the contractor for poor construction, and the latter blames the design team for poor design; and
- the process is slower than other procurement methods because detailed design needs to be completed before tender and construction.

Design and build

The common dispute between the project team and contractor as well as poor project performance led to the development of the design and build (DB) delivery method. Here, the design team is separated from the project team and integrated with the contractor (Figure 6.5). This provides the owner with a single point of responsibility. The contractor cannot blame the designer for poor design, and the latter cannot accuse the former for shoddy work.

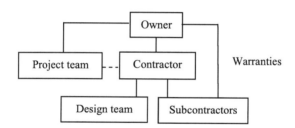

Figure 6.5 Design and build method.

Based on the owner's project brief, pre-selected DB contractors are asked to propose an outline design and formal price proposal. At this stage, the role of project team, which now comprises primarily the project manager and cost engineer (quantity surveyor), is to advise the owner on the design brief, tender documentation, cost, and selection of DB contractor.

There are many variations in the DB delivery method. Since construction work is highly cyclical, many contractors do not have in-house specialist designers to avoid being saddled with large fixed costs. If the owner perceives that the contractor's in-house design team is not strong, she may engage a design team first to control design and then novate it to the contractor during the construction stage. A disadvantage with novation is that the design team no longer represents the owner's interest. However, this is true only if the project is one-off; in repeated projects ("games"), the design team needs to serve both masters if it wants future work.

By having a single point of responsibility, the owner can expect faster project delivery. Detailed design need not be completed before construction, and the squabble between the designer and contractor is eliminated or internalized. This is important if projects need to be launched to time the property cycle or save on the huge interest bill.

Turnkey contract

A turnkey contract is an extension (or variation) of a design and build contract. It includes procurement, design, construction, installation of equipment, and commissioning. In some cases, it may also be extended to include feasibility studies, land acquisition, short and long-term financing, licensing, technical assistance, operations, maintenance, and training of operation staff members (e.g. on how to operate complex machinery). The product is a fully-equipped and functional facility with a contractor's performance warranty. Hence, the client needs only to "turn the key" at the end of the project.

A turnkey contract is used for specialized work where the contractor (or a consortium) is able to provide or procure the full complement of services listed above. Terms such as limited or partial turnkey contracts are also sometimes used to indicate various combinations.

Turnkey contracts retain the advantage of the design and build delivery method of a single source of responsibility and faster project completion. However, the client may lose control over decision-making as many aspects of the project are dealt with by the contractor. In addition, cost control is harder if there is no detailed cost estimate. Finally, the turnkey contractor must be experienced with the technology to minimize design, construction, and installation problems. If the technology is overly sophisticated, the client may be too reliant on the contractor for licensing, repairs, maintenance, upgrades, and training.

Management contracting

Management contracting is a fee-based delivery system where the management contractor uses his expertise to manage the design and construction stages for a fee and a guaranteed maximum price (GMP) for the project. If the project cost is below GMP, there is often provision for sharing the difference (e.g. 50:50). Clearly, if the contractor focuses on the split, construction quality may suffer. GMP need not be used solely for a fee-based management contract; it is possible for a contractor to combine a cost-plus-fee approach with a guaranteed maximum price.

The design team may be engaged by the owner (Figure 6.6) or by the management contractor (shown dotted). In the former arrangement, the design team is independent of the management contractor. Both the design team and management contractor work together to design and build the project, and early contractor input allows fast-tracking of the project and fewer design changes. Nonetheless, disputes between the two parties are not uncommon.

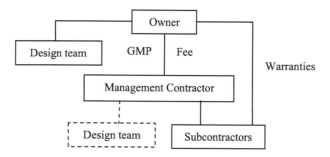

Figure 6.6 Management contracting.

Project financing model

The project financing model is an entirely different method of procurement. Its main characteristic is the setting up of a special project vehicle (SPV) by project sponsors to handle the feasibility study, design, construction, and post-construction management (operation) of the facility (Figure 6.7). Sponsors and other passive investors contribute equity to the SPV and the rest of the funds are borrowed on non-recourse or limited-recourse basis. This means that lenders rely primarily on project assets and cash flows to service the debt. This model is used to finance risky and capital-intensive projects.

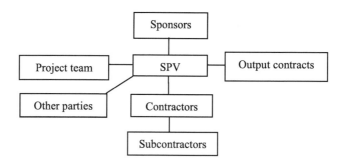

Figure 6.7 Project financing model.

The public sector may be an equity partner. Alternatively, public goods and services may be procured using a public-private partnership (PPP) or public finance initiative (PFI). Here, a public agency enters a PPP contract with an SPV to rent, for example, a hospital built and operated by the SPV. For more details, see Chapter 8.

6.9 Design development

Regardless of the procurement method selected, the client needs to have some control over the design of the facility. For instance, in the traditional lump sum delivery method, the architect completes the design before contractors are invited to tender for the project. In this case, the client has considerable control over the design, and the contractor provides minimal input to design.

In design and build (DB) projects, the design is evolving to fast-track the project. Based on the RFP, short-listed DB contractors are invited to present their design and formal price proposals. Once a design has been accepted by the owner and the DB contractor has been appointed, the design develops from conceptual to detailed design.

Periodically, the owner will review the design with her project team and contractor. Note that the price has been fixed or guaranteed at this early stage of a DB project. The benefit for the owner is early knowledge of cost and doing away with the need for detailed drawings, specifications, and pricing at the initial stage where time is critical. For the contractor, mispricing without detailed drawings is a major risk. Hence, a client should select an experienced contractor; mispricing by a contractor may be bad news for the client as well.

The owner and architect will usually identify a few design issues. Examples include

- the need to design the facility in such a way that subsequent construction work will minimize disruption to existing tenants;
- the desire to preserve existing rare trees;
- integration of the development with an existing market or feature (e.g. underpass); and
- the desire to create a particular type of atmosphere (e.g. festive).

Based on the client's requirements in the form of a design brief, the architect produces a land use plan and conceptual designs (schematic diagrams) for approval based on the number of units. For instance, in a residential development,

Gross Floor Area (GFA) = Plot Ratio × Site Area.
Net Floor Area (NFA) = Building efficiency × GFA
Number of units = NFA/Average unit size

The building efficiency is about 0.85 so that NFA is merely GFA less ("dead") spaces that are not sellable (e.g. staircases).

Once the conceptual scheme has been approved by the owner, the design enters the second stage. The owner will provide a second brief on the detailed requirements and specifications of the components or elements of a building or facility.

Thereafter, the architects (some architectural works are likely to be subcontracted to specialist designers, e.g. design of swimming pool) and engineers proceed to design the development in greater detail. Each building will be divided into separate elements that later form the work breakdown structure. An example is given below.

Substructure
>   Piling
>   Foundation
>   Excavation
>   Basement
>   Waterproofing

Superstructure
>   Frame
>   Roof
>   External walls
>   Windows
>   External doors
>   Internal walls
>   Internal doors

Internal finishes
>   Wall finishes
>   Floor finishes
>   Ceiling finishes

Fittings and furniture

Services
>   Sanitary
>   Drainage
>   Water supply
>   Heating
>   Air-conditioning
>   Ventilation
>   Fire protection
>   Connection to utilities

External works
>   Club house
>   Car park
>   Internal roads
>   Footpaths
>   Fencing
>   Landscaping
>   Sports facilities
>   Pool
>   Playground

As the design proceeds, the dimensions of each element are measured by the quantity surveyor and transferred into Bills of Quantities (BQ). Along the way, the quantity surveyor will continue to update and refine initial estimates of the project cost.

6.10 Detailed estimates

In the traditional delivery method, the BQ is first priced by the consultant quantity surveyor on the client's side using unit rates from price manuals, experience, or market feedback. Some clients do away with this step of detailed pricing and shift price risk to the contractor. In such cases, the client is experienced enough to know the approximate price and this price is capped either through tender or a Guaranteed Maximum Price.

The unpriced BQ, together with drawings and contract documents, is used for tender to be priced quantity surveyors hired by contractors. The prices from both sides are then compared for various purposes such as to detect measurement errors, omissions, suicide bids, negotiate with contractors, and so on. The BQ contains the measured elements as well as the following:

- Preambles that define the quality of materials and workmanship;
- Provisional sums for work where information is incomplete such as when detailed design has not been completed;
- Prime cost sums attributed to nominated subcontractors; and
- Contingencies.

Subcontractors may not estimate their bids to that level of detail. For instance, a small work package may be estimated as shown in Table 6.3. Often, a simple checklist is used to avoid errors and omissions.

Costs	$
Labor	
Skilled −100 hrs @$50/h	5,000
Unskilled − 200 hrs @$20/h	4,000
Material	10,000
Equipment	5,000
Others (e.g. Transport)	1,000
Total labor and non-labor costs	25,000
Overhead@10%	2,500
Subtotal	27,500
Contingency@10%	2,750
Total	30,250

Table 6.3 Cost estimate for small work package.

6.11 Tender documents

In the traditional delivery method, a construction tender contains the following standard documents:

- Letter of invitation to tender;
- Conditions of tender specifying the time and place for meeting with project team and client, tender deposit (or bid bond to ensure the winning

bidder signs the construction contract and not withdraw because of high workload or mispricing), closing date for bids, location to submit bids, method of award, expected date of award, and expected start date;
- General conditions of contract that establishes the legal responsibilities of parties to the contract, obligations, authority, and rights;
- Special conditions of contract such as additional insurance required from the contractor apart from worker's compensation and "All Risks" insurance (inclusive of third party insurance);
- Contract form, which is usually a standard form of contract that specifies the parties, description of work, contract sum, start and end dates, liquidated damages, method of progress payment, interest for late payment, retention, and final payment;
- Drawings (architectural, mechanical and electrical, civil and structural, and so on);
- Specifications including dimensions, materials, method of construction, standard of workmanship, and performance standards;
- Bills of Quantities for contractors to apply unit rates and estimate tender price;
- Bid forms specifying the name of contractor, tender price, price breakdown for major trades, amount of bond, alternates (prices of alternative materials or method of construction), fees for additional work recommended by contractor, and nominated subcontractors; and
- Addenda of changes made before bids are due.

6.12 Contractor selection

The selection of designers (i.e. architects and engineers) is usually based on a combination of track record, fees, conceptual design, and previous working relations.

Contractors are often selected using suitable weights on items such as track record and expertise (including safety), financial strength, workload, bid price, and schedule. The weights are may be based on judgment (e.g. 10 per cent for track record and expertise, and so on) with greater weight on bid price. In more complex projects, value management may be used to derive the weights as shown in the example below.

Example

Suppose the criteria are:

A: Track record and expertise
B: Financial strength
C: Workload
D: Bid price
E: Schedule

Criteria	A	B	C	D	E	Frequency	Weight
A		A	C	D	A	2	0.2
B	A		B	D	E	1	0.1
C	C	B		D	E	1	0.1
D	D	D	D		D	4	0.4
E	A	E	E	D		2	0.2
Total						10	1.0

Table 6.4 Derivation of weights.

Criteria	Weight	Contractor			Contractor		
		X	Y	Z	X	Y	Z
		Original score			Weighted score		
A	0.2	6	7	8	1.2	1.4	1.6
B	0.1	5	6	9	0.5	0.6	0.9
C	0.1	9	6	5	0.9	0.6	0.5
D	0.4	6	8	7	2.4	3.2	2.8
E	0.2	6	8	5	1.2	1.6	1.0
Total	1.0				6.2	7.4	6.8
Bid ($m)					100	105	108
Value ratio					0.062	0.070	0.063

Table 6.5 Selection of contractor based on value ratio.

In Table 6.4, comparisons are made between pairs of criteria by a decision panel. For instance, in the first row, A is considered to be more important than B, less important than C and D, and more important than E. The row result is then transposed to the column by symmetry. In the second row, B is more important than C but less important than D or E. Again, the results are transposed to the column. For each row, the frequency of the criterion is then noted from which weights are derived. A criterion with zero frequency is dropped from consideration.

The weights are then transferred to the second column of Table 6.5. Columns 3 to 5 show the original scores for each of the three contractors (X, Y and Z) on each criterion on a 1-10 rating scale with 10 as the best score. The scores are then multiplied by the weights and totaled in the last three columns. Based on value ratios, contractor Y should be selected.

Some procurement procedures do not allow for contract negotiation with the lowest bidder. In many cases, clarifications on contract terms and specifications are required and this may allow scope or opportunity for price "negotiation." Playing off contractors to drive down bids is generally not acceptable but it does happen in practice. If bids are close, two or three short-listed bidders may be invited to submit revised "best and final offers" taking into account revised specifications, drawings or contract terms. If the client is a public agency, significant changes will normally lead to reopening of a new round of bidding competition.

There are many other variations on how contractors may be selected. In a design and build project, design may be evaluated separately from price rather than integrated in the example above. If the design and other qualitative aspects are evaluated on a score of 0 to 100 (expressed as a decimal), then a bid is adjusted by dividing the raw bid by the design score (e.g. ($100 m)/0.85 = $117.6 m). In other cases, the owner may stipulate and fix the contract sum and the best design is then selected.

6.13 Construction

Whatever the delivery method, the main concern during construction is to achieve the triple objectives of delivering the project on time at a reasonable cost and quality. Scope creep, cost escalation and quality deterioration are key challenges. In addition, the project manger

- tracks progress;
- manages project data;
- manages the risks;
- manages design changes (variation orders and change orders) using an appropriate control and approval system; and
- communicates with key stakeholders on project status and relevant issues.

The workhorse in project scheduling is the Gantt chart. Since details may be found in standard texts in project management, only a brief description is given here using a simple example (Table 6.6).

Task	Resource			Duration (Quarters)					
	Workers	Eqpt	Cost ($m)	1	2	3	4	5	6
A	10	2	5	▬					
B	15	2	10		▬	▬			
C	5	2	3		▬	▬	⌐ ⌐		
D	5	4	4				▬	⌐ ⌐	
E	20	4	30				▬		
F	3	1	2					⌐ ⌐	
G	20	1	10						▬
			Resource						
			Workers	10	20	20	25	20	20
			Eqpt	2	4	4	8	4	1
			Cost	5	30	30	50	30	10

Table 6.6 Gantt chart.

Columns 2 to 4 show the resources required for each task. The duration of each task is plotted as horizontal bar charts. Some activities (e.g. foundation) need to be completed before other activities can proceed. For instance, task *A* needs to be completed before task *B* can start. A thick horizontal line shows the progress of a task; for example, tasks *A* and *B* have been completed.

The next step is to determine the critical path on which any delay would delay the entire project. The critical path is shaded with dots, that is, tasks *A*, *B*, *E* and *G*. For simple projects, this path may be identified visually, and arrow diagrams may be used if simple heuristics or visual methods are impractical. Since the Gantt is merely a planning tool, it is also important not to neglect "near critical paths" where delays will lengthen project duration.

Activities not on the critical path may be floated, and the extent of float is shown in dotted lines. For instance, tasks *C* and *D* have one quarter of float (or slack) each.

The vertical line denotes "time now" and shows that the project is currently in the middle of the 4th quarter. Task *C* is behind schedule but since it is not on the critical path, it is not necessary to speed (crash) it up by using more resources. The activities to crash depends on

- project progress;
- whether a task is on the critical path or near-critical path; and
- resources available; and
- relative costs and penalties.

The relative cost depends on the cost gradient (Figure 6.8) and penalty for late delivery. For task *G*, the cost gradient is ($1 m)/2 = $0.5 m per month. The gradient for task *E* is ($2 m)/3 = $0.67 m per month. Hence, it is cheaper to crash task *G* if the amount spent is less than the corresponding penalty for delay.

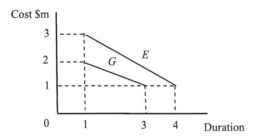

Figure 6.8 Cost gradients.

The Gantt chart may be used for resource leveling by changing the resource profile at the bottom of Table 6.6. The resource profile is obtained by adding each column vertically for each resource used for that quarter. Since 25 workers and 8 pieces of key equipment (e.g. cranes) are needed in the 4th month but only 20 workers and 4 pieces are needed in the 5th quarter, starting task F later in the 5th quarter will even out the resource profile.

From the Gantt chart, it is also possible to compute the cost-duration and work-duration curves (Figure 6.9). In the top panel, the actual project cost expended at "time now" is below the planned cost. The bottom panel shows the project is ahead of schedule.

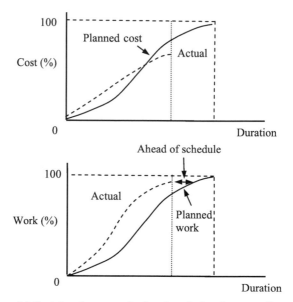

Figure 6.9 Cost-duration curve (top) and work-duration curve (bottom).

As the construction proceeds, the owner may need to change certain aspects of the design or because of changes in regulation. The project manager or architect will originate a Change Proposal Request (CPR), typically a one page form, to the contractor. The CPR identifies the change, parties involved (or effected), price, and expected duration. Each request is then recorded in the Change Order Log.

Upon receipt of the CPR form, the contractor provides the requested information to the owner for her decision to proceed with the change, modify, or cancel it.

If the owner decides to proceed with the change, the CPR becomes a change order, and a form, signed by the three parties, is used for this purpose. The Change Order form authorizes the contractor to carry out the work and obligates the owner to pay for it. Depending on circumstances, there may or may not be an extension of project time.

Generally, change orders put owners at a disadvantage, and contractors may use it as an opportunity to overcharge or recoup losses from a low tender. On the other hand, a contractor may not necessarily agree with a substantial change order (e.g. if it considerably delays a project and he is committed to another project). In such cases, the contract is likely to oblige the contractor to proceed with the change directive. It may end in a dispute.

Note that change orders may also change the value of a contract. Hence, consent from the surety for the contractor's performance bond is required. The additional bond premium to cover the changes is normally paid by the originator of the change (i.e. owner).

Project managers should understand that a change order is a contract between the owner and contractor. All too common, overzealous project managers tend to start "ordering" the contractor around without realizing the seriousness of a change order. If a contractor consented to a verbal simple change "order," project managers should never assume that the contractor would not bill it later and ask for extension of time as well.

6.14 Project close-out

There are two main activities at project close-out. One is the takeover procedure that includes commissioning or user acceptance tests (UAT) of the performance of various facilities, as-built drawings, and where appropriate, user training and manuals (e.g. software). In construction projects, there is a retention sum (about 5 per cent) to ensure the contractor rectifies defects within a year.

The second main activity is the project audit report that includes the following:

- Executive summary;
- Project review of objectives and approach;
- Performance of project team in terms of time, cost and quality;
- Project deliverables and assessment;
- Lessons learned;
- Individual team member assessment and recommendations; and
- Overall recommendations.

Questions

1   Recall from Equation (6.1) that

$$\text{NPV} = -C_0 + \frac{N_1}{1+r} + \frac{N_2}{(1+r)^2} + \cdots + \frac{N_n}{(1+r)^n}.$$

Show that if N_t and r are in *nominal* terms, NPV is unaffected only if revenues and costs per period are identically affected by inflation.
[Hint: Use the relation $1 + r = (1 + \text{real discount rate})(1 + \pi)$]

Note: In practice, all revenues, costs, and rates of return are reckoned in nominal terms on the assumption that inflation effects on costs and revenues are identical.

2   The NPV curves for projects *A* and *B* are shown below. Explain why the use of
    NPV (at discount rate *c*) and project IRR criteria leads to different ranking of
    projects.

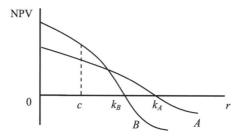

3   The NPV curve below shows that there are multiple IRRs, that is, there is more
    than one root to the equation NPV = 0. Under what circumstances are multiple
    roots possible?

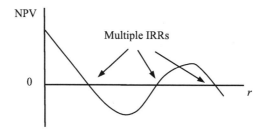

4   Your friend offers a sale and leaseback arrangement with you on a laptop. He
    will sell you his laptop for $2,700 cash and lease it back from you at $1,000
    per year payable at the end of each year for 3 years. All repairs will be borne
    by your friend, and at the end of 3 years, he will buy back your laptop for
    $200.

    a) What is the NPV if the discount rate is 5 per cent? [$196]
    b) What is the project IRR? [8.7%]

5   After realizing that the project IRR is 8.7 per cent in Question 4, you manage
    to find a loan of $2,000 at 4 per cent annual interest which you will pay off by
    the end of 3 years.

    a) What is the annual repayment for the loan? [$721]
    b) Check that the amortization table below is correct and then use it to compute
       the cash flows. [$F_1 = \$279; F_2 = \$279; F_3 = \$479$]
    c) What is your equity IRR? [20.3%]

Year (a)	Principal at start of period (b) ($)	Annual repayment (c) ($)	Payment breakdown	
			Interest (d) ($)	Principal (e) ($)
1	2,000	721	80	641
2	1,359	721	54	667
3	692	721	29	692
4	0			

Social Projects

7.1 Private and social considerations

Infrastructure projects undertaken by the State on a social rather than private basis may be planned and executed in a top-down manner through central planning of the investment requirements of each sector (as in socialist countries) or bottom-up through benefit-cost analysis (BCA) of individual projects. BCA is also called cost-benefit analysis or social cost-benefit analysis. In the latter, a project is worth undertaking if it results in net social benefit (i.e. social benefits exceed social costs) or, equivalently, there is an increase in social welfare.

The bottom-up project approach is nowadays more popular with governments, international lending and aid agencies because of greater control over expenditures and intended outcomes. First applied in the 1930s in water resources projects in the United States, BCA is now the standard tool for evaluating social infrastructure projects.

Sometimes, BCA is viewed too narrowly merely as a financial tool. As we shall see below, BCA is quite versatile and goes beyond efficiency and project financial considerations. Distributional issues and the valuation of intangible benefits and costs also form part of BCA. In some cases, it may not be possible to identify or value the intangible benefits and, as a result, cost effectiveness analysis (i.e. cost minimization) is used to assess projects.

For infrastructure and other social projects undertaken by the State, the private costs and revenues discussed in the previous chapter may not fully capture the social benefits and costs generated by the project. Inevitably, there will be winners and losers, and project evaluation is inherently political. For example, a dam that provides city water will flood smaller towns and villages along the river and damage parts of the ecosystem or historical sites. A new highway that bypasses a small town will affect businesses and employment in that town.

Not all "losers" are adequately compensated. In theory, the improvement in social welfare is only potential, that is, if winners can potentially compensate losers and still have some positive net benefit. This so-called Kaldor-Hicks criterion is unsatisfactory, and BCA has been rightly criticized for it. In practice, losers should as far as possible be adequately compensated in one form of another. For instance, residents and businesses in flooded towns and villages should be given alternative accommodation and premises. Damage to the environment should be properly valued and minimized, and steps should be undertaken for replanting or rejuvenation of forests. If possible, historical sites should be excavated to save the artifacts for future generations.

Like private projects, social projects also contain risks such as incorrect demand projections, cash flow problems, construction problems, and operational difficulties. In addition, there may be "government failure" to achieve efficiency and equity. Examples of such failures include corruption, monopoly, inefficiency due to lack of competition, inadequate supervision, insufficient skills or resources to execute too many ambitious projects, and poor coordination or turf politics among State agencies.

Recall from Equation (6.2) that, for *private* projects,

$$\text{NPV} = -C_0 + \frac{N_1}{1+k} + \frac{N_2}{(1+k)^2} + \cdots + \frac{N_n}{(1+k)^n} \tag{7.1}$$

where C_0 is initial project cost, N_t is the net operating income in year t, n is project duration, and k is the project internal rate of return (IRR). To compute the equity IRR, Equation (6.3) is used.

In considering *social* projects using BCA, the net operating income is replaced with net social benefit (i.e. social benefit (B) less social cost (C)) so that

$$0 = (B_0 - C_0) + \frac{B_1 - C_1}{1+s} + \frac{B_2 - C_2}{(1+s)^2} + \cdots + \frac{B_n - C_n}{(1+s)^n} \tag{7.2}$$

where s is the social project IRR or, if equity and cash flows are used instead of initial cost and net operating incomes, s is the social equity IRR.

The essential difference between private and social projects is not whether social project or equity IRR should be used but how to value and assess the distribution of social benefits and costs. In the private sector, profit is simply revenue less cost using market prices. However, the use of market prices to determine social costs and revenues underplays

- social benefits based on willingness to pay;
- social costs that are not captured by the market;
- the distinction between real and pecuniary effects;
- price distortions due to monopoly (including labor unionization), taxes, subsidies, tariffs, and so on;
- unregulated externalities or third-party costs and benefits such as damage to the environment or negative impacts on residents (e.g. noise) without adequate compensation;
- the option value of natural areas;
- the distributional impact of projects on different social groups;
- project sustainability;
- multiplier effects of large infrastructural projects;
- the interests of future generations through an appropriate social discount rate; and
- income effects.

These issues are discussed below. In essence, they are related to well known "market failures" of price distortion, missing markets, externalities, and imperfect information.

It is worth noting that, like private sector project evaluation, BCA also tries to reduce social benefits and costs to a single dimension, in dollars. While this allows the ranking of projects based on IRR, it should be remembered that BCA is not a purely technical undertaking. Its aim is to improve decision-making in evaluating projects, not to replace subjective judgments.

7.2 Valuation of social benefit

Consider the demand curve (D) for a private good or service such as a cup of coffee in Figure 7.1. Each point *on* the demand curve indicates the *maximum* amount of money consumers are willing to pay (WTP) for the benefit of consuming each marginal cup of coffee. For example, the first consumer is willing to pay 90 cents for a cup of coffee, the 50th consumer is willing to pay 60 cents, and the 100th consumer is only willing to pay 30 cents.

If the unregulated market price based on the intersection of the demand and supply curves is 30 cents, then all consumers pay this price. Since the 50th consumer is willing to pay 60 cents, he enjoys a "surplus" of $60 - 30 = 30$ cents. That is, coffee is viewed personally as relatively "cheap" to this consumer.

Adding up the surpluses of each consumer who is willing to pay higher than 30 cents, we obtain the (total) consumers' surplus given by the area of the triangle $A + B + C$.

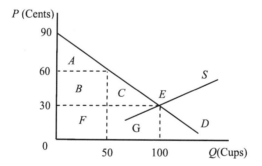

Figure 7.1 Valuation of social benefit.

In the private sector, the consumers' surplus is usually ignored, although nonlinear pricing may be used under certain conditions to capture it. For instance, movie goers who value watching a popular film early may be charged higher prices than those who can afford to wait. Such complications, while obviously important in some practical cases, are not considered in this chapter.

In the private market, the seller sells each cup of coffee for 30 cents, and obtains a total revenue of 100 × 0.30 = \$30 (= $F + G$). Thus, *the market underestimates the social benefit or total value of a good or service to society.* The social benefit is the area under the demand curve, that is, $A + B + C + F + G$, which is greater than the revenue received by the seller (i.e. $F + G$).

7.3 Valuation of social cost

On the cost side, the supply curve (S) represents the additional (marginal) cost of producing each extra unit in a competitive industry. If the demand curve shifts outwards (such as due to a rise in income or tax fall), the price rises from p to p^* and producers are willing to supply more units (from q to q^*) because it is profitable (Figure 7.2).

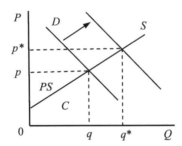

Figure 7.2 The supply curve.

Let $Q = f(K, L)$ be the production function where Q is net output per period (net of materials cost, i.e. Q represents value-added), K is capital input (inclusive of land input) per period, and L is labor input per period (e.g. a year). It is understood that K and L are *flows* of services per unit time from existing *stocks* of capital and labor. For instance, L is measured in man-hours per year (a flow concept), not the number of workers (a stock concept). Similar, a stock of capital equipment yields a flow of capital services per unit time. The cost of this flow of capital services is the rental rate of capital.

In the *short run*, one factor is assumed to be fixed (e.g. K) so that $Q = f(L)$ only and successive application of additional labor leads to diminishing returns and hence rising marginal cost. Thus, the short-run supply curve (S) is upward-sloping. A simple example is the cost of building new homes. If prices rise because of new demand (such as from lower interest rates), developers will bid for the limited amount of land in the short run. Hence, land costs will rise. Similarly, there may not be sufficient skilled construction workers in the short run to cope with the new demand, and wages rise. Alternatively, if wages remain constant, developers are using less skilled workers, and productivity falls. The effect is the same: marginal costs rise.

In the *long run*, both K and L are not fixed and the long-run supply curve may be upward-sloping, horizontal, or downward-sloping depending if there are decreasing, constant, or increasing returns to scale respectively. That is, if inputs are all increased by the same factor λ (e.g. if $\lambda = 2$, all inputs are doubled), the new output, with technology held constant, is

$Q^* = f(\lambda K, \lambda L) < \lambda Q$ (decreasing returns to scale)
$\quad\quad = \lambda Q$ (constant returns to scale)
$\quad\quad > \lambda Q$ (increasing returns to scale)

In other words, if all inputs are doubled and output is less than double, we have decreasing returns to scale. If doubling inputs double output, there are constant returns to scale. Finally, if output is more than double, we have increasing returns to scale. Importantly, note that technology is held constant; otherwise, it is not possible to distinguish pure scale effects (i.e. returns to scale) from effects of technical progress in increasing output.

If p is the equilibrium price determined by the intersection of the demand and supply curves, suppliers are willing to supply output q at total cost C. The revenue received is $0p \times 0q$ or $C + PS$. Just as the market gives consumers a surplus, the area *PS* is called the producers' surplus.

Example

If \$2 m worth of equipment plus 10 workers can build 6 houses a year, how many houses will be built if both inputs are doubled?

To answer this question, we need to know something about house-building management, geometry, finance, and technology. If the houses are next to each other, there may be increasing returns to scale if equipment and workers can be employed more efficiently and if grouping them together does not lead to higher industrial disputes. If they are spatially separated, then coordination problems and transport costs rise, and there may be decreasing returns to scale.

Geometry has scale effects as well. If a cylinder has radius r and height h, its volume is $\pi r^2 h$. If the radius and height are doubled, the new volume is $\pi(2r)^2(2h) = 8\pi r^2 h$, which is more than twice the original volume. Similarly, if the basic dimensions of a house are doubled, the volume of the expanded space is more than twice the original one.

On the financial side, by borrowing more funds, the firm may face rising interest cost to compensate lenders for higher default risk. On the other hand, it may lead to lower interest rates as well because of lower transaction costs.

Finally, physical laws may also have scale effects. If the dimensions of a cylinder are doubled, the walls may need to be thickened, but by not as much. This is a physical property of the strength of materials.

The concept of returns to scale assumes inputs are scalable up or down (i.e. divisible). This is rarely true for most capital goods. For instance, computers cannot be scaled down proportionately; neither can cars. Some parts are scalable; some parts simply cannot be scaled proportionately without adversely affecting product performance. In most cases, returns to scale operate only approximately.

7.4 Real and pecuniary effects

Since BCA deals with social welfare, it is generally concerned with real effects directly in terms of resources used (land, labor, and capital) and indirectly in terms of social benefits and costs such as pollution, the flooding of homes in the case of a dam, or destruction of the environment.

In contrast, pecuniary effects or pure financial transfers are ignored (Prest and Turvey, 1965). If a new road raises the profitability of a petrol station and reduces another that has been bypassed, such pure transfers are ignored on the ground that no real resources are used up in the process. Similarly, if a firm relocates from one area to another because of the road project, employment is transferred from one area to another.

In practice, the firm may make use of the opportunity to build a better plant, use a new production process or hire more workers. These effects are no longer pecuniary.

7.5 Incremental outputs

If the project produces a *new* product or service, then the benefits and costs are straightforward, as shown below.

In Figure 7.3, D is the demand curve and S is the supply curve which is assumed to be horizontal to simplify the exposition. If a new product or service is produced and priced at p, the total benefit to consumers is $A + X + Y$, and the total cost is $X + Y$. Hence, the net benefit of the project to per unit time (e.g. each year) is $A + X + Y - (X + Y) = A$.

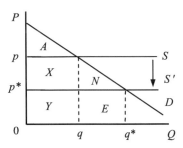

Figure 7.3 Incremental output.

Consider a different and common situation where it is an *existing* product or service and the project produces an increment to the existing output. One example would be a power project where the current capacity is q (Figure 7.3) and the upgraded capacity is q^*. As a result of the increased capacity, the price of electricity falls from p to p^*. The *change* in benefit to *existing* consumers as a result of the price fall is X, but this is exactly offset by the loss of X from existing producers.

For *new* consumers, the total benefit is the area $N + E$ and the cost is E. Hence, the net benefit to new consumers is N.

In summary, there is a transfer of surplus (X) from existing producers to existing customers and the net social gain for these two groups is zero. However, for new consumers, the net benefit from the new power station is N. Hence, the net gain from the power project to society each year is N. In terms of Equation (7.2), we can write the numerator for any year t as

$$N_t = B_t - C_t.$$

For simplicity, N may be computed for the first year of operation and a growth factor is applied for each subsequent year.

Often, it is assumed that the demand curve in Figure 7.3 is linear so that N is estimated as the area of the triangle, that is,

$$N = \tfrac{1}{2}\,(p - p^*)(q^* - q) = \tfrac{1}{2}\,\Delta p \Delta q. \qquad (7.3)$$

Hence, one needs to estimate the price change (Δp) and the incremental output (Δq). This is a reason why economists are very interested in the price elasticity of demand given by

$$\varepsilon = (\Delta q/q) \div (\Delta p/p).$$

Since Δq is harder to estimate than the observed change in prices (Δp), an alternative formula to Equation (7.3) may be derived by eliminating Δq using the two equations above to obtain

$$N = \tfrac{1}{2}\,\varepsilon q(\Delta p)^2/p. \qquad (7.4)$$

If it is assumed that a one per cent fall in the price of electricity leads to a one per cent rise in consumption, then $\varepsilon = 1$. Conventionally, the minus sign is ignored in price elasticity so that $\varepsilon = 1$ instead of -1. Since most econometric estimates of the long-run value of ε (e.g. Bohi (1981)) are close to one, another approximation to Equation (7.3) is

$$N = \tfrac{1}{2}q(\Delta p)^2/p.$$

Here, only the current price of electricity, expected price change, and current consumption are required to estimate N.

7.6 Price distortions

Output and input prices may be "distorted" by price controls, subsidies, taxes, tariffs, overvalued exchange rates, and monopolistic pricing. For instance, land prices may be distorted by zoning, rent control, tradition (e.g. communal lands are not for sale), and compulsory land acquisition. Such distortions, which can be severe in less developed countries, are neglected in the profit calculus of private sector entrepreneurs.

In evaluating social projects, "distorted" prices need to be "corrected" to reflect real resource value. If market prices are also missing, shadow or proxy prices need to be computed.

On efficiency grounds, it has been argued that the market maximizes the net social benefit. In terms of Figure 7.4, this means that, at the equilibrium price p, the sum of producers' surplus (*PS* or area of triangle *pEF*) and consumers' surplus (*CS* or area of triangle *GEp*) is maximized. An alternative way of looking at it is to consider an output (*m*) at less than equilibrium output q. Then the marginal social benefit (point *a*) exceeds the marginal social cost (point *b*). Society benefits by increasing output from *m* to q.

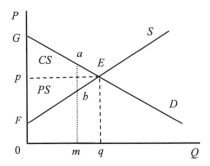

Figure 7.4 Efficiency of the market.

Consider what happens if price is "distorted" by a price control or collusion that fixes the price (p^*) above the market equilibrium price (Figure 7.5). The consumers' surplus has shrunk to area *GHp**, and the producers' surplus is now *p*HJF*. The sum of consumers' and producers' surplus is *GHJF*, which is less than *GEF*, the sum of surpluses in a competitive market. As a whole, society (i.e. consumers and producers) loses the area of the triangle *HEJ*, called the deadweight loss because nobody gains it.

The allocation is said to be Pareto inefficient because there are losers (consumers). An allocation is Pareto efficient if it is not possible to make anyone better off without making someone worse off. Note that the Paretian criterion is a value judgment. If a cake is to be shared between two parties (*A* and *B*), *any* division is Pareto efficient because changing it will make one party worse off. However, it need not be equitable. For instance, if *A* gets 99 per cent of the cake and B gets 1 per cent, the allocation is Pareto efficient since any re-division towards B will make A less well off. But not many people will think the 99%:1% split is equitable, particularly if the "cake" is housing or income.

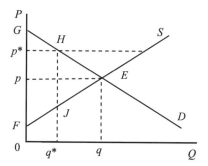

Figure 7.5 Price distortion and inefficiency.

Apart from the static efficiency discussed above, the market is also dynamic. The profit motive and competition compel firms to continually learn from experience (i.e. learning by doing; see Solow (1997)), innovate, and cut costs. Hence, the supply curve does not stay still; it may be shifting downwards in response to cost-cutting measures and upwards in response to adverse supply shocks.

So the free market is generally more efficient than if prices are "distorted." Is the free market equitable? The neoclassical answer is based on marginal productivity theory (Borjas, 2004). Workers (or any other factor of production such as capital and land) will not be hired at less than the value of their marginal product or, equivalently, their marginal contribution to output. Since all factors of production are paid according to their individual contributions to output, the distribution of that output is said to be equitable. There is no free lunch except for those who are too young or physically unable to work.

The above partial equilibrium framework based on the workings of a specific market does not constitute a proof of general market efficiency or equity. In lieu of a proof, Adam Smith argued long ago for the existence of an "invisible hand" to run the market system but it is only a metaphor.

The proof that a general equilibrium exists in a competitive economy is given by Arrow and Debreu (1954). Even here the restrictive assumptions should be noted, namely, voluntary exchange, rational individuals, self-interest, and perfect information.

In reality, information is imperfect or asymmetrically held by different parties such as the lender and borrower. Further, rationality is bounded by the ability of the mind to process the large amount of information, resulting in the search for "satisfactory" rather than optimal solutions (Simon, 1957). The result is differing expectations about the future and this contributes to market instability.

7.7 Wage distortions

Wages as prices of labor may be distorted in the labor market by minimum wage legislation, trade union practices, entry barriers, discrimination, and payment of "efficiency wages" above market level to attract better workers (Akerlof and Yellen, 1986). For instance, remuneration in certain professions may not fall with the influx of new university graduates because of stringent post-graduation entry barriers.

In addition, if workers are drawn from the rural sector, there will be a difference between agricultural wages, the urban informal sector wages, and industrial wages. There is some dispute over which level of wage reflects the real cost of hiring additional labor for the project. These issues are discussed in the next section.

7.8 Shadow prices

Given that a project's input and output market prices may be distorted, how may shadow prices be estimated to reflect real opportunity costs? There are two approaches, namely, the Little-Mirrless (1974) or World Bank method, and the UNIDO (1972) approach. The UNIDO approach is more complex and hence less popular. It will not be considered here.

Little and Mirrlees proposed that, for countries with large price distortions, world *import* prices of tradable goods net of all tariffs and taxes may be used instead of controlled prices and then converted to domestic prices using a "suitable exchange rate." For example, consider the case of an imported item:

Exchange rate	1US$ = 2 pesos
Border price	20 pesos (US$10)
Tariff	10 pesos
Local transport	3 pesos
Distribution	2 pesos
Total market price	35 pesos

The shadow price is 25 pesos (i.e. it excludes the tariff). Put differently, one can apply a conversion factor to the market price (35 pesos) to obtain its shadow price (25 pesos).

If the project output is *exported*, a similar logic applies except that an export tax, if it exists, it added to the factory gate price and local transport price to obtain the shadow price at the border (or free on board (FOB) price) just before the product is shipped. Recall that for imported inputs, the tariff is excluded in computing the shadow price. The export tax is not excluded in the case of exports because it is not treated as a transfer for the purpose of evaluating the feasibility of a project.

It is not easy to estimate a "suitable exchange rate" to convert border prices to shadow prices. In practice, the official exchange rate is used because it is easily available and its use is not questioned in official feasibility study reports. However, black market exchange rate may be much higher.

If a good is not tradable (e.g. local transportation and utilities), it is broken down into its tradable and non-tradable components and world prices may also be applied to tradable parts. The non-tradable components are broken down again into tradable and non-tradable components, and so on. The process is complex and tedious, and specific conversion factors used in converting distorted market prices into shadow prices are only rough estimates.

For the adjustment of distorted wages into shadow wages, the industrial wage should be used as the shadow wage since workers need to be compensated for the higher cost of living, travel, pollution, and productivity. Surplus rural workers should not be assumed to be unemployed or under-employed with zero opportunity cost (e.g. Lewis, 1954) or that the labor supply curve bends backwards because unemployed rural workers prefer leisure.

Some analysts recommend using the informal wage (Harberger, 1972) to reflect the low productivity of these workers working in the informal (bazaar) sector, the average rural wage, the casual rural wage, or a weighted average of different wages (Bruce, 1976). In practice, rural wages are either not properly reported or highly seasonal and hence not easy to determine. Thus, there are theoretical and practical reasons for using the industrial wage.

7.9 Externalities and missing markets

Externalities are third-party costs and benefits. In infrastructure projects, examples of negative externalities include the destruction of the natural environment and pollution. On the other hand, vaccination of children provides a positive externality by reducing the number of deaths and the risk of a disease spreading to the general public. Similarly, the benefit of education accrues to the individual as well as society in having better educated people.

A public park provides social benefits and yet there is no market for such parks. This is because it is difficult or not possible to exclude anyone from using a public park by imposing fees. If fees cannot be charged because of non-exclusion, the private sector will not provide the public good and, without market prices, it is difficult to value its benefit.

Similarly, the cost of destruction of the natural environment is not easy to ascertain. Contingent valuation (Mitchell and Carson, 1989) based on household surveys is sometimes used to value the environment. Respondents are asked how much they are willing to pay to support measures to save the environment or how much compensation they require should it be destroyed. Such surveys are prone to large errors because it is difficult to put a value to it in the first place, the questions are hypothetical, respondents may answer out of self-interest or interviewers may bias the responses.

For recreation areas such as a nature reserve, it may be possible to use the travel cost method (Clawson and Knetsch, 1966) to derive the demand curve. An example is given below.

Example

Suppose the area around a nature reserve is divided into three zones (1, 2 and 3). The data in Table 7.1 gives the annual visits from each zone if the admission fee is zero. It is then possible to plot the last two columns and draw the line of best fit (Figure 7.6).

If the admission fee is raised to $1, we have the data in Table 7.2. Recall that if there is no admission fee, there are 6,000 visits. With the $1 admission fee, the number of visits is 4,610. This gives us two points on the graph in Figure 7.7. The procedure is repeated by setting admission fee to $2 and working out the number of visits to find the third point. The curve joining these three points is, of course, the demand curve.

Zone	Population	Visits	Visits per 1,000 population (V)	Transport cost per visit ($)
1	8,000	4,000	500	2
2	3,000	1,000	333	4
3	4,000	1,000	250	6
Total		6,000		

Table 7.1 Data on annual visits.

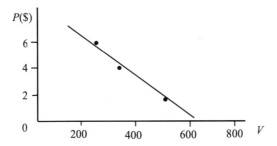

Figure 7.6 Plot of transport cost and V.

Zone	Population	Transport cost + fee ($)	Visits per 1,000 population (V) estimated from regression line	Total visits (TV)
1	8,000	3	400	3,200 (= 400 × 8)
2	3,000	5	230	690 (= 230 × 3)
3	4,000	7	180	720 (= 180 × 4)
Total				4,610

Table 7.2 Annual visits if admission fee is $1.

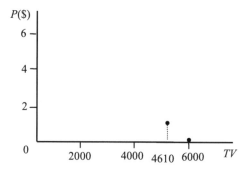

Figure 7.7 Plot of admission fee against *TV*.

Problems with the travel cost method include possibly biased survey data on transport cost, arbitrary delineation of zones, different types of visitors (e.g. those on holidays as opposed to nearby residents), whether people have the time to visit such places and, perhaps most important, how it is valued by non-visitors. Generally, attempts to measure environmental benefits and costs are fraught with difficulties, and the way forward may be to treat them as political rather than economic variables.

The impact of externalities may also be measured through the use property values. If we have a cross-sectional sample of *n* houses, then the price of each house is given by the hedonic price function

$$p_i = \beta_0 + \beta_1 x_1 + \cdots + \beta_k x_k + \varepsilon_i \qquad (7.5)$$

where β_0 is the constant term or intercept, $\beta_1,\ldots,\beta_k$ are parameters to be estimated using the sample of n $(> k)$ houses, $x_1,\ldots,x_k$ are housing characteristics (land area, location, tenure, amenities, and so on), and $\varepsilon_i \sim N(0, \sigma^2)$ is the normally distributed error term with zero mean and constant variance. The parameters may be estimated using ordinary least squares (OLS).

Hypothetical data for the estimation are shown in Table 7.3. For example, the first house is sold for $1.2 m, is 1.9 km from the Central Business District (CBD), has a swimming pool, and so on. The variable "Pool" is called a dummy or categorical variable; it takes the value "1" if the house has a pool and "0" otherwise.

House	Price ($'000) P_i	Distance from CBD (km) x_1	Pool x_2	Etc
1	1,200	1.9	1	
2	900	2.3	0	
...				
200	1,400	1.5	1	

Table 7.3 Data from sample of 200 houses.

A possible regression result is

$$E[P_i] = 300 - 20x_1 + 10x_2 + \cdots + 5x_k \qquad (7.6)$$

where $E[.]$ is the expectation. On average, each kilometer away from the CBD leads to a \$20,000 *fall* in house prices. Similarly, on average, each pool *adds* \$10,000 to house price, and so on.

The use of hedonic price models in valuing social benefits or costs is now apparent. Suppose we wish to value the impact of noise from a new highway. Although there is no explicit market for traffic noise, there are implicit prices because noise is capitalized into property values. Hence, a possible so-called hedonic price model is

$$P_i = \beta_0 + \beta_1 x_1 + \cdots + \beta_j \text{NOISE} + \cdots + \beta_k x_k + \varepsilon_i \qquad (7.7)$$

where NOISE is a dummy variable. It is set to "1" if a house is next to a busy road and "0" otherwise.

The hedonic price model postulates that the price of house may be decomposed into individual contributions by a bundle of house characteristics. Since prices for these characteristics are implicit rather than explicit (only house price is observable), the markets for these characteristics are called implicit markets.

If a large sample of houses located next to a major road or major roads is available, it is possible to estimate the values of each of the parameters. The estimated coefficient for NOISE is likely to be negative because of the negative impact of noise on house value, all else equal. For instance, if the estimated value of β_j is -10, then each house located next to a busy road is, on average, worth \$10,000 less, holding other variables constant. In other words, if there are two identical houses apart from location, then the one impacted by traffic noise is worth \$10,000 less on average.

Summing over all houses affected by the proposed highway, we obtain the total loss in house values due to traffic noise. This sum is then converted to annual values by multiplying it by r_h, the rate of return to housing investment in the area (= 0.05 say). The logic is that if a house priced at $\$P$ is net rented at $\$R$ annually in perpetuity, the rate of return is $r_h = R/P$. Hence, the annualized value, R, is given by $\$r_h P$.

Hedonic pricing is also used to measure the adverse impact of airport noise, air pollution, proximity to toxic waste, and so on. The main limitations of the model are well known and summarized below (see Tan, 2006):

- adverting behavior – the house owner may install double-glazed windows to ameliorate the effect. This feature of the house may not be captured as one of the independent variables;
- market failure – the error term is often quite large, not only because of measurement and omission errors but also because the housing market may not be working properly. Hence, if the effect of air pollution is found to be negligible, it may be because the market does not capture this effect, not because air pollution has no effect on property values;

- market instability – if the market is unstable, house prices tend to change sharply within a short period of time, and the estimated coefficients are unstable; and
- small sample size – since a large number of coefficients need to be estimated, the sample size should be large. However, given the need to collect data on house prices during only stable periods, there may be few transactions.

Given these limitations, it is not surprising that the accuracy of hedonic prices is rather low. At best, they provide a rough estimate of the impact of a project on property values.

A positive externality when a road is improved is the number of lives saved from road fatalities. The value of life may be estimated using the human capital method of discounting foregone future earnings to present value. From the average age of road accident deaths, one may estimate the average future earnings of someone of that age and then discount it to present value using a reasonable discount rate. A flaw with this method is that many road accident victims are the young and very old.

7.10 Option value

The destruction of nature generally has irreversible consequences and this gives rise to the option value or willingness to pay for the option of using it in future (Krutilla, 1967). This option value is not priced by the market and hence cannot be captured by private enterprise.

In addition to option value, it is sometimes argued that a nature reserve has existence value (or sentimental value), that is, one places a value on its existence even though there is no intention to use it in future.

Both values are difficult to determine and they are either neglected in BCA or left to decision-makers to decide.

7.11 Distributional issues

When considering private projects, the entrepreneur does not bother with distributional issues. However, a social project that benefits only the rich and impacts negatively on the poor is unacceptable since actual compensation may not be paid to those adversely affected by the project.

A theoretically simple solution to the distributional problem is to use "social pricing" rather than economic pricing by giving more weight to the utility values of the poor but it is impractical and the proposal has never gone past theory (Little and Mirrless, 1990; Squire, 1989). Utility cannot be measured in "utils" and therefore cannot be summed, let alone weighted and maximized as social welfare for the greatest good.

A more realistic approach is to identify the benefits and costs of the project to different groups of stakeholders (Jenkins, 1997). Project impact should not be an after thought; rather, it should be incorporated in the design stage to target certain beneficiaries, particularly the poor.

7.12 Project sustainability

Project sustainability relates to issues such as

- environmental sustainability or the balance between meeting the needs of current and future generations with respect to the environment. A key concept is environmental stress from resource depletion, pollution, and destruction of ecosystems. Such stresses may be measured using an index of land, water, and air quality, extent of waste, and biodiversity;
- institutional capability to deal with the human and environmental stresses caused by the project and such stresses may spill over international borders;
- economic sustainability or the absence of major macroeconomic imbalances; and
- financial sustainability in terms of funding adequacy and cost recovery.

7.13 Multiplier effects and development

It is well known that projects have direct and indirect multiplier effects. For instance, a major road project creates new employment directly for surplus agricultural labor (Lewis, 1955), and these incomes earned by workers are spent on other goods and services, thereby creating several rounds of multiplier effects in the economy.

The project will also directly benefit suppliers, subcontractors, lenders, land owners, lawyers, professionals, and insurers on the input side. On the output side, the car, oil, tourism, and related industries benefit from higher demand for vehicles and travel.

Even the State benefits through higher tax revenues from rising profits, incomes, land values, property values, stamp duties, and payments for regulatory services.

Beyond multiplier effects, an intercity road project is also an instrument of economic development by creating new industries, addressing the urban-rural imbalance in resource allocation (Lipton, 1977), and overcoming the failure of markets to coordinate private decisions by encouraging the setting up of complementary industries to extend the limits of the market and reduce the risk of setting up businesses (Nurkse, 1953).

Generally, only the first one or two rounds of multiplier effects are considered in BCA. It is difficult to trace subsequent rounds of multiplier effects without the use of extensive input-output tables showing the linkages among industries in the economy (Leontieff, 1986). However, these detailed tables are expensive to produce, and many countries either publish them every five to ten years or do away with it completely. For these reasons, input-output tables will not be considered here.

7.14 Choice of hurdle rate

Since discounting is used, the choice of hurdle rate is important in evaluating projects. If a high discount or hurdle rate is used, many social projects will not be feasible.

The discount rate for social projects reflects the interests of current and future generations. A society has a preference to consume now or in future. If wealth is not spent, it may be saved by buying a bond at real risk-free interest rate r_f. If S_0 is saved in period zero, the amount of savings in the next period is

$$S_1 = (1 + r_f)S_0.$$

The riskless after-tax real interest rate, assuming zero transaction cost, is also called the social rate of time preference. It is often taken to be about 4 per cent per year.

If part of the savings is invested by firms and Q_1 is output in the next period, then

$$Q_1 = (1 + r_s)Q_0,$$

where r_s is called the social opportunity cost of capital. In real terms, it is often taken to be about 8 per cent per year after tax.

If there are no taxes or investment externalities, the two rates (r_f and r_s) are equal and given by the intersection of the demand and supply curves for loanable funds (Figure 7.8). If a tax is imposed on savings, the supply curve shifts downwards by the amount of the tax to S^* and the two rates diverge.

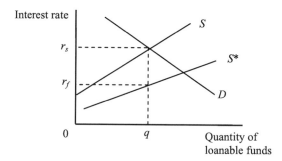

Figure 7.8 Choice of discount rate.

Which discount rate should one use to evaluate social projects? One preference is to use the opportunity cost of capital. It is argued that if a public project cannot earn a better rate of return than a private project, it should not proceed by diverting funds from the private sector.

Supporters of a lower hurdle rate for social projects (i.e. r_f) point to market "myopia" to prefer present to future consumption. Consequently, less weight is given to the preferences of future generations. In addition, they point to the substantial benefits of social projects (e.g. multiplier effects). However, using too low a social discount rate has two dangers: misallocation of resources, and possible damage to the environment.

A compromise is to use the weighted average (Sandmo and Dreze, 1971; see also Diamond, 1968)

$$r = \theta r_f + (1 - \theta)r_s$$

where θ is the proportion of public investment obtained by foregone private investment. If r_f is 4 per cent, r_s is 8 per cent, and θ is 0.2, then the social rate of discount in real terms is

$$r = 0.2(4\%) + 0.8(8\%) = 7.2\%.$$

Thus, if the rate of inflation is 1 per cent, the nominal social hurdle rate is 8.2 per cent.

7.15 Income effects

The discussion so far assumes that if a project causes a change in prices, it does not affect the incomes of consumers. For instance, consider the demand and supply of transport services in Figure 7.9. P_0 is the original price determined by the intersection of demand and supply curves. If the existing road is upgraded, the supply curve S_0 falls to S_1 and P_0 falls to P_1. The quantity of road services demanded rises from Q_0 to Q_1.

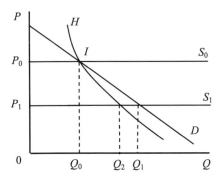

Figure 7.9 Hicksian demand curve.

The price fall changes the real incomes of road users. Recall that shifts in the demand curve are due to non-price changes such as demography, interest rates, income, and tastes. Hence, the price fall that affects real incomes will shift the demand curve. It is no longer correct to use the original demand curve to compute the area of the triangle in Equation (7.4).

To overcome this problem, we need to estimate the location of the new demand curve. Starting from initial position I, we remove part of the income from users so that they will purchase slightly less road services Q_2 ($< Q_1$). The resulting demand curve H is called the Hicksian or compensated demand curve.

Usually, the price changes are small so that the income effect is negligible. Hence, in most practical cases, the original uncompensated demand curve (D) rather than the Hicksian demand curve is used (Willig, 1976).

7.16 Case study

BCA can be quite complex. The brief case study here is intended to illustrate how the various sections in this chapter fit together.

Suppose there is a proposal to build a dam. The possible benefits and costs are given in Table 7.4.

	$m
Initial costs	
Land	50
Construction	200
Plant and equipment	10
Destruction of wildlife and forest	Unpriced
First year benefit	
New agricultural land @$100/ha	20
Recreational uses	2
Flood reduction	Unpriced
Water for consumption	10
First year costs	
Loan repayment	10
Operating and administrative costs	5
Maintenance and repairs	1

Table 7.4 Example of BCA.

The value of new agricultural land may be estimated from the rural land rent market provided land prices are competitive and there are sufficient land transactions to discover prices. If agricultural products are subsidized, land rents no longer reflect competitive market values and these values must be adjusted to market levels.

The initial costs for construction and plant and equipment are assumed to be reasonably accurate. The destruction of wildlife and forest (as well as flood reduction) is left unpriced for politicians to make the decision whether the project should go ahead.

The benefit from recreational uses for the first year may be estimated using the travel cost method. As shown in Figure 7.10, there will be q^* visitors if there is no admission charge and the social benefit is given by areas $A + B + C$. If a fee (p) is imposed, the number of visits falls to q and the social benefit is given by area $A + B$.

The benefit from water consumption for the year is based on Figure 7.11. Water from the dam shifts the vertical supply curve outwards, resulting in a fall in water prices from p to p_1 and a consequent rise in water consumption from q to q_1. Existing consumers gain by area A but this is exactly offset by the loss of revenue to existing water producers. The total benefit to new consumers is $B + C$ and net benefit is $B + C - B = B$. Hence, the net social benefit to existing and new consumers is B.

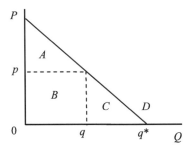

Figure 7.10 Benefit from recreational use.

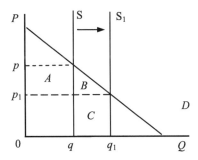

Figure 7.11 Benefit from water consumption.

There are many possibilities such as the new dam will displace some producers or traditional sources of village water supply (wells, rivers, and so on). These complications are ignored here.

The first year costs are straightforward. Once the benefits and costs for the first year are estimated, a constant growth rate may be applied to obtain the annual benefits and costs for subsequent years. These net benefits are then discounted to compute the IRR.

Questions

1 Explain why the use of market prices and quantities in BCA underestimates the social benefit generated by a project.

2 Explain why the use of market prices and quantities may also underestimate the social cost.

3 For the good below, value the annual
   (a) social benefit; $[A + B + C]$
   (b) social cost; and $[C]$
   (c) net benefit. $[A + B]$

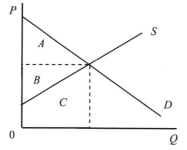

4 The current cost of one-way air travel between two cities is $200 and there are one million passenger-trips. If air travel is deregulated, the price is estimated to fall to $150 and passenger-trips are expected to rise to 1.3 million. Calculate the annual net benefit of deregulation. [$7.5m]

5 If the savings rate is 4 per cent, the return to capital is 12 per cent, and 20 per cent of total investment consists of private investment, compute the social discount rate. [5.6%]

6 It is sometimes argued that project managers should assess project viability using objective facts, leaving the questions of values for politicians to decide. Equivalently, project managers should concern themselves with the means, and let politicians determine the ends. This is the so-called fact-value or means-end distinction. Is this position defensible, and why?

7 The record for long-term projections for rail projects in the US is not good. Pickrell (1989) studied ten cases and found that the capital and operating costs

were often badly under-estimated, and ridership was grossly over-estimated. What explains these results?

8   It is sometimes argued that discounting places very little weight on events that occur in distant future, and future generations may suffer from the way we value finite resources in favor of the current generation. For instance, $1,000 m 100 years from now is only worth $1,000/(1.05)^{100} = \$7.6$ m today if the discount rate is 5 per cent. Evaluate this claim.

9   On the opening day of a new highway that bypasses a few congested towns, a politician remarked that the new highway would boost local businesses, generate employment, and raise incomes for residents. Evaluate this claim from the perspective of benefit-cost analysis.

10   The term "soft budget" constraint was coined by Kornai (1986) from his observations in Hungary that unprofitable State-owned firms were not allowed to fail. The State would bail them out in times of financial crises through capital injections, tax cuts, or restructuring of debt and this, according to Kornai, was a major cause of inefficiency. An important issue is whether a soft budget constraint may operate in non-socialist settings (see Question 7), in social projects. Discuss.

8

Characteristics of Project Finance

8.1 The structure of project finance

In this chapter, we consider the contractual relations among project stakeholders with direct or indirect interests in the project. A general model of the structure of project finance is shown in Figure 8.1.

There are many variations to this basic model. For instance, bonds may not be issued in a project, and lenders may include international lending agencies such as the World Bank and Asian Development Bank. Further, each party may assume several roles. For example, in addition to its regulatory role, the State may also be the supplier of inputs (e.g. oil) as well as the off-take purchaser of the project output through another agency such as the State Electricity Board in the case of power projects.

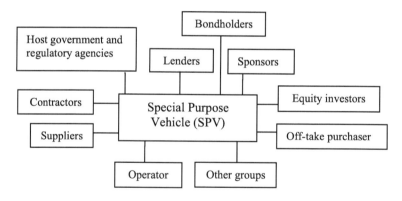

Figure 8.1 The structure of project finance.

Recall from Chapter 1 that project finance is a way of financing a capital-intensive project on non-recourse or limited-recourse basis through a legal entity or special project vehicle (SPV). One of the project sponsors may also operate the facility.

In pure non-recourse project financing, only project assets and cash flows are used for loan repayment, and this makes non-recourse lending risky. Lenders are often compensated by the opportunity to lend substantial sums of money on lucrative projects. In addition, lenders impose various loan covenants on the SPV to manage risks and enhance project success.

In limited recourse lending, parent companies of project sponsors provide some form of contingent financial support over and above their equity share as well as other forms of credit enhancements and third-party guarantees.

The financing is structured through debt and equity. In many cases, the requirement for large sums of money necessitates lending by a syndicate of banks led by a lead manager or arranger. If additional funds are required, layered or mezzanine finance with repayment priority lying between senior debt lenders and equity investors is also used.

We saw in Chapters 4 and 5 why project financing might be attractive to the firm. Debt is cheaper than equity because of tax deductions, and the use of higher debt implies little equity is required to finance a risky project. Further, debt can be kept off the balance sheet of the parent company through an SPV to allow the firm to maintain its credit standing and continue to borrow from other lenders for its other activities. The project risk is also separated from the firm's overall level of risk, and therefore does not adversely affect the latter's share price. This separation of risk also leads to lower lender screening cost (Shah and Thakor, 1987). However, there is some loss of control for the parent firm (Chemmanur and John, 1996).

8.2 Corporate finance

The structure of project finance in Figure 8.1 may be contrasted with conventional corporate finance (Figure 8.2).

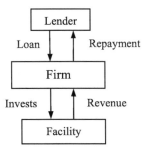

Figure 8.2 Conventional corporate finance.

The firm uses debt and equity to finance the project. Lenders have full recourse to the assets of the firm if it defaults on its loan repayments. Further, the debt shows up on the firm's balance sheet. This may not be desirable if leverage is already high, affecting investors' valuation of the firm's share price and its ability to borrow additional funds.

On the positive side, corporate finance has a simple structure as well as easier and cheaper to arrange. Since lenders have full recourse to corporate assets, loan covenants are also less stringent.

8.3 Conventional public procurement structure

The project finance structure in Figure 8.1 may also be contrasted with the traditional or conventional public procurement structure in Figure 8.3.

In the past, governments tended to fund, build, and manage their own facilities. The funds may come from tax revenues, foreign aid, international lending agencies, domestic lenders (provided they are keen to lend to such projects), issuance of State bonds, or domestic savings through a State-owned bank, housing funds, pension funds or postal savings schemes.

In many cases, even under the traditional procurement method, the design, construction, operation, and maintenance were carried out by the private sector under different contracts.

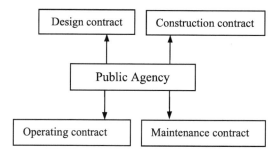

Figure 8.3 Conventional public procurement structure.

This conventional model does not tap the full range of expertise and funds from the private sector. In addition, ownership of the facility ties down limited State funds. These are some of the reasons why many governments nowadays are reluctant to use the traditional approach and more keen on public private partnerships, which are discussed next.

8.4 Public-private partnerships (PPP)

Following the fiscal crises of the 1980s in many countries and the swing towards "market friendly" policies promulgated by the World Bank and International

Monetary Fund to get "prices and policies right," many governments sought to tap the large amount of funds, resources, and expertise of the private sector.

One way of achieving this objective is through public-private partnerships (PPP) (Figure 8.4) or public finance initiative (PFI). Consequently, with encouragement from the World Bank, the financing structure to procure public facilities began to shift from the traditional public procurement system towards the project financing model in Figure 8.1. For critics, PPP was seen as a form of "back door" privatization, but in the end, it made sense not to tie up substantial State funds in bricks and mortar. PPP is sometimes misunderstood as a panacea for project failures. There are many causes of project failure from initiation to completion. The form of financing is certainly a major factor, but it is not the only cause of failure.

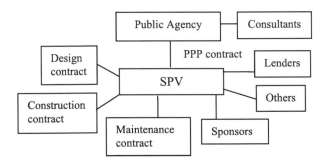

Figure 8.4 PPP procurement structure.

In a PPP, the public agency identifies the need, scope, timelines, (single or unitary) price, and output service levels using key performance indicators. A team of consultants may be appointed to assist the public agency to prepare the items above.

Often, the public agency is able to provide the site and invites the private sector to participate in the project. A Request for Qualification (RFQ) is used to pre-qualify bidders. After a reasonable period (e.g. a month) for shortlisted bidders to provide feedback and seek clarification, bidders are asked to present their proposed designs and bids based on the guidelines given in the Request for Proposal or RFP (see Chapter 6 for details on RFP). If successful, the private sector participants form an SPV to finance, build, operate, and transfer (if appropriate) the project after about 20 to 30 years.

The exact delivery method varies from project to project. The transfer may take effect by having a site lease of only about 20 to 30 years. In the case of road projects, the SPV may be able to cover the costs by imposing agreed tolls. In cases where the SPV is likely to incur a deficit, the public agency may pay an annual fee. The SPV maintains the facility, leaving operational matters (e.g. running of a public hospital) to the public agency.

If the builder is also the operator, he will take into consideration the life cycle of the facility during design and construction to minimize subsequent maintenance and operation costs. This is a major advantage of a build-operate-transfer or build-operate-own procurement system.

8.5 Stakeholders

Stakeholders are individuals, groups or organizations with direct or indirect interests in the project. They may be categorized as

- internal stakeholders within the SPV such as sponsors, the project team, and workers; or
- external stakeholders as shown in Figure 8.1 as well as parties such as insurers, affected businesses, landowners, the mass media, environmental activists, and residents.

Brief notes on stakeholders are given below.

8.6 Sponsors

Sponsors are organizations, corporations or individuals that establish the SPV for the purpose of financing and executing the project. A consortium of sponsors may be set up to pool resources (particularly equity and complementary expertise) and share the risks.

Generally, sponsors hold only a small share in the SPV. For instance, if 30 per cent equity is required and there are three sponsors and two passive equity investors, then each sponsor's share is only 6 per cent if equity contributions are equal for each party. This means that, for accounting and legal purposes, the SPV is not considered to be a subsidiary.

In some cases, the State is also a sponsor with equity investment to show commitment as well as the ability to acquire land and expedite potential regulatory bottlenecks.

Sponsors provide the business case, conduct feasibility studies, and promote the project to lenders, governments, and the public. They also play an active role in running the project through the Board of Directors and management team of the SPV as well as in managing other stakeholders.

8.7 Equity investors

Since projects are capital-intensive, sponsors may not have sufficient equity or wish to limit their equity exposure. The remaining equity is supplied by passive equity investors who are paid periodic dividends and do not interfere with the management of the project.

Passive equity investors may be institutions such as investment trusts, insurance companies, pension funds, trade union funds, or wealthy individuals.

Unlike many project sponsors who operate within an industry, passive equity investors generally own a portfolio of assets. Hence, they are better able to diversify specific project risks.

8.8 Host government

The host government may play a number of direct and indirect roles, and these are outlined below. Governments have different capacities to perform these roles. The list below merely states the possible roles a government can play; whether they actually perform such roles is specific to each project. More than one host government may be involved in a project. For instance, an oil pipeline may traverse different countries, and different governments are usually involved in such a strategic project.

Political roles

- promoter of the project concerned with providing a good or service (e.g. a new highway) at an affordable price;
- garner political support for the project; or
- mitigate political risks such as civil unrest, expropriation of assets and status of agreements following a change of government before project completion.

Macroeconomic roles

- enhance macroeconomic stability (national debt, inflation, interest rates, output, and employment);
- ensure convertibility of currency and repatriation of profits;
- ensure adequate investment in physical and social infrastructure; or
- build institutions for growth and development, such as the development of financial markets.

Microeconomic roles

- develop land and labor markets;
- correct market failure including the creation of "missing markets" such as particular types of insurance and protection of the environment; or
- develop fair competition policies.

Regulation and law

- effective management of permits and permissions;
- regulate contractors and subcontractors;

- promote fair competition;
- enforce contracts; or
- protect property rights.

Financing

- participate as equity partner in projects of strategic or national interest;
- guarantee loan repayment if borrower is a State agency;
- use foreign aid effectively; or
- provide tax incentives.

Inputs

- acquire land through powers of eminent domain; or
- guarantee input supplies if it is a supplier.

Operations

- provide licenses;
- provide appropriate investment incentives;
- provide fair utility rates; or
- other concessions to attract investment.

Output

- guarantee payment of output if the buyer is a government-linked firm;
- periodic or one-time payment (i.e. contingent contributions) to the SPV for the shortfall in revenue below the guaranteed minimum; or
- provide non-monetary contributions such as granting sponsors the right to develop adjoining land.

Businesses are often in two minds about State participation. On one hand, if the State performs the roles outlined above well, it will benefit businesses substantially. On the other hand, State participation can lead to fiscal stress, corruption, false reporting, lack of transparency, poor coordination, expropriation of assets, uncertainties over regulatory requirements and decisions, weaknesses in enforcing property rights and contracts, poor commitment, and insufficient checks and balances.

States should not be simply labeled as effective or ineffective. Detail analysis is required to understand the nature of the problems. Indeed, "the State" is also a constellation of group and individual interests as well as institutions and they differ across jurisdictions.

In the context of project finance, of particular importance is fiscal stress at the different State levels. In many countries, the distribution of taxing powers is uneven, leading to vertical imbalances between revenues and expenditures at different levels of government. There are also horizontal imbalances at the same level of government such as between urban and rural municipalities. With fiscal decentralization, the general principle is for the central government to tax income to finance its expenditure, leaving local jurisdictions to tax immobile property to finance local public goods more suited to variations in local tastes (Oates, 1972). Non-public goods and services should be charged rather than financed out of general taxation.

This logic of fiscal decentralization and revenue sharing is supposed to make governments accountable and responsive to national as well as local pressures. Critics argue that, in practice, local States are poor implementers of projects because of the dominance of certain interest groups and weak institutional capacity. Hence, outcomes of fiscal decentralization are contextual (Bird and Vaillancourt, 1998).

On the expenditure side, there is a general contraction in welfare payments in "post-welfare States" but a rise in spending on items such as military and anti-terrorism, education, medical services, infrastructure, and housing. Hence, there tends to be a structural gap or mismatch between State expenditures and revenues, leading to fiscal stress and even periodic crises.

The problem is particularly acute at local levels of government. Property taxes are often sufficient to cover only operating expenses. Capital expenditure needs to be supported by central government aid, grants, or transfers. In many cases, local government borrowings are required to cover the shortfall but investors may not view local governments as credit worthy or responsible. In projects where cost recovery is limited, such as water supply and sanitation projects, it is difficult to attract private capital.

8.9 Lenders

There are various types of lenders in a project, and the main ones are discussed below.

Construction lenders often comprise a syndicate of banks that provides short-term construction and land loans for commercial projects or long-term loans for infrastructure projects. A syndicate of banks spreads the risk among lenders and also helps raise the substantial amount of funds required. International agencies such as the World Bank, International Finance Corporation (IFC), European Investment Bank, and Asian Development Bank (ADB) are often construction lenders in infrastructure projects.

Permanent lenders are required in commercial property projects to "take-out" the short-term loan from the construction lender. The permanent lender may comprise institutional investors such as Real Estate Investment Trusts (REITs) and insurance companies. Figure 8.5 shows how a large mall or office development is financed based on development stages rather than the contractual relations among parties in Figure 8.1. The REIT may be an active managing equity investor, that is, it manages the facility for a fee in terms of marketing, rent collection, maintenance, and so on directly or through a subsidiary. Alternatively, the facility may be securitized and sold to investors to "unlock" its value and provide the trust with the liquidity to expand its portfolio of assets.

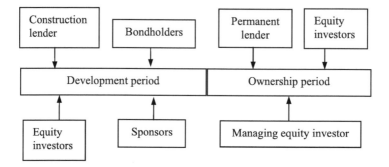

Figure 8.5 Possible financial arrangement for a property project.

Equipment lenders or suppliers such as export credit agencies provide loans for the import or purchase of equipment.

Bondholders subscribe to the bonds issued by SPV and are paid periodic dividends. In the past, bonds were seldom used during the construction phase because of the perceived project risks or because of underdeveloped domestic bond markets. However, the trend is changing; with better risk management and higher yields, bonds may be attractive to investors, particularly if they are issued when a project is nearing completion so as to remove a substantial part of the construction risk. Indirectly, the State may also issue tax-exempt bonds to finance infrastructure projects.

8.10 Suppliers

There are many types of suppliers, and the main ones include equipment, raw materials, fuel, and utility suppliers.

Suppliers expect to be paid on time and, where exotic materials are used, sufficient time should be provided for order and delivery. Prototypes and samples cost money to procure and experiment. Lastly, suppliers do not expect to be blamed for monopolistic practices for sharp rises in prices, such as that of iron, steel, and oil in 2004 and 2005.

8.11 Contractors and consultants

Contractors and consultants are involved in the engineering (design), procurement, and construction (EPC) of the facility. They include the main contractor, subcontractors, designers (engineers, architects, interior designers, and other specialists), project managers, cost engineers or quantity surveyors, and other professionals.

Project managers are concerned with meeting the triple constraints of time, cost and quality. In trying to meet these objectives, they constantly have to coordinate and negotiate with other professionals over design, construction, and quality issues.

Consultant quantity surveyors or cost engineers fear mispricing from incorrect measurements, omissions, or using incorrect unit rates amid aggressive tender deadlines. For the responsibilities and effort in measuring to such detail, they wish fees could be higher.

Generally, architects focus on aesthetics and like to be proud of their work. Budget and schedules feature less in their minds, unless they are waiting to be paid for their effort. Good architects produce their completed drawings with clarity and on time.

Engineers are more concerned with the civil, structural, mechanical, and electrical aspects of the facility. Professional liability, safety, and being paid on time feature highly.

Contractors want to be paid on time so that the credit chain to suppliers, workers, subcontractors, and other bills is not broken. In projects where profit margins are reasonable, they tend to accommodate requests, provide free services, and appreciate referrals. In less profitable projects, there is little room for extras and plenty of space for disagreements.

Subcontractors share the contractor's concern over cash flow. They are often treated as if they are at the end of the value chain. If a project falters financially, they tend to be the unhappy casualties. If contractors are not paid, subcontractors can expect to be squeezed.

8.12 Operator

The operator operates and maintains the facility after it has been built. In build-operate-transfer projects, the operator will transfer the facility to the owner after the operation contract expires.

There are variations such as the build-operate-maintain-transfer arrangement where maintenance is the responsibility of sponsors and the build-own-operate structure where the facility is not transferred to the host government. As noted earlier, the integration of design, build, and operations provide incentives for the design/build team to incorporate operational and maintenance issues during design and construction.

8.13 Off-take purchaser

Since projects are risky, lenders may require that there is an off-take purchaser of the project output. In power projects, the off-taker is usually the local electricity provider. In petrochemical projects, the off-takers are committed upstream chemical firms.

In office and retail projects, lenders require pre-leased anchor tenants to enhance project success. Typically, about half of the commercial spaces need to secure pre-leasing agreements.

Not all projects have off-takers. For instance, road projects do not have ready "purchasers" of road services based on long-term contracts. The viability of such projects is based on projected demand.

8.14 Other stakeholders

The other stakeholders of a project include workers, insurers, affected businesses, landowners, residents, the mass media, and other interest groups with direct or indirect interests in the project.

8.15 Stakeholder politics

Beyond knowing the structure of project finance and the roles and responsibilities of each stakeholder, project managers require a basic understanding of stakeholder politics to manage projects effectively.

A critical question in stakeholder politics is to determine who decides on key issues. One naïve answer, which is no longer popular, is that "capital decides" and shapes the rural and urban landscape. The answer is too simple because it ignores the role of local politics. For this reason, we need to examine the nature of local politics a little further.

Dahl (1961) has argued that local politics, at least in New Haven in the US, is fragmented. In this *pluralist model*, many large-scale community projects were initiated by the mayor and "sold" to the community. People from all walks of life participated in decision-making, the business leaders were mostly passive, and there was no dominant social class. However, critics have argued that what is perhaps more important is that sensitive issues are kept off the agenda, leaving public debate to relative safe matters. In turn, this critique assumes that the public are unaware of the main issues, and this position may not stand up to empirical scrutiny.

A more common view of local or city politics is that decisions on key project issues are determined by the *power elite* led by business leaders and politicians (Hunter, 1953; Miliband, 1969). On one hand, it is impractical for any outside investor to negotiate with a large group of people on many issues. On the other hand, the power elite is also interested in competing for external investment to boost economic growth, property values, and create jobs.

This power elite is sometimes called a "growth machine" in city politics (Molotch, 1976). The machine metaphor underscores the efficiency as well as the dominance of the pro-growth coalition in local politics. However, not all city leaders are pro-growth promoters, which means the concept of a "growth machine" is not well theorized. Hence, the "growth machine" may itself be divided internally depending on the issue.

For this reason, Stone (1993) has argued that, depending on what is at stake, "*urban regimes*" of politicians, top civil servants, and business leaders formed pro-growth coalitions to get things done. For these so-called "political entrepreneurs," coalition-building is an ongoing process of establishing and enhancing social networks and dominating the ideological apparatuses such as schools, religious institutions, and the mass media. These efforts require resources and political connections that only members of the regime can afford or have access. On minor matters or "safe" issues, the power elite takes a back seat and allows for public participation.

The main weakness of both the pluralist and elite models is the absence of the social and economic contexts that constrain the actions of political actors (Poulantzas, 1969). Politics is not an autonomous sphere, and it is only "relatively autonomous" from the material conditions of the economy, an idea that may be traced back to Marx (1963) in his examination of the rise to power of Louis Bonaparte as a temporary but independent force.

What, then, are the mechanisms that constrain the local power elite, so that it is unable to do what it likes?

Business leaders within the power elite are constrained by State institutions in terms of its agencies, regulations, and rules. The State need not merely react to the motives and actions of community leaders. Instead, it has its own interests in securing legitimacy of its rule as well as supporting the accumulation of capital to finance State expenditures (O'Connor, 1973). The latter include social investment to assist accumulation of capital (e.g. investment in research and development, education, housing, health, and physical infrastructure), and social expenses to legitimize its rule (e.g. expenditure in the military, police, and welfare payments).

In turn, the State is also constrained by the ballot box, popular uprisings, and the mobility of capital in raising revenue. Although capital is taxable, it is highly mobile and capitalism develops unevenly across regions. Hence, the State acts to attract capital through low taxes and subsidies, and different States or regions compete among themselves for investible funds. This mobility of capital is a major constraint on State action. If the State over-regulates production (e.g. by imposing stringent environmental or labor requirements), capital can withdraw from the place in the long run. The balance of power between the State and capital differs across places. In declining regions, States are desperate for investment and will tend to offer many concessions. Conversely, in booming regions, States can afford to be more stringent.

Within the State, the actions of the local State are also constrained by the central or federal State (in a federated system) through budgetary or financial controls. The local State may not have a free hand in considering projects for approval if it wants federal financial assistance. However, as discussed earlier, the federal government is also constrained by its own budget. In times of fiscal stress, the urban, city, or regional "problem" may not be seen a federal issue at all. In the new free market ideology since the 1980s, regions, cities, and municipalities must "look after themselves" through a series of self-help initiatives rather than rely too much on federal assistance. In place of its interventionist role, the federal government tends to see itself as facilitator or enabler.

The local State is also constrained by revenues from local property tax. Tiebout (1956) has long argued that residents can "vote with their feet" by shifting from one local jurisdiction to another to avoid paying high property taxes without a commensurate level of local services. Indeed, local States are keen to attract good businesses and high tax-paying residents and repel pollutive industries and the urban poor who consume local urban services and do not pay much tax. Several well-known mechanisms for doing this include the "not in my backyard" mentality to ward off undesirable industries, large-lot zoning, and under-zoning "unsuitable" land uses (e.g. public housing).

In summary, the State does not have full autonomy to do what it likes as if it is an independent force. It is constrained by the economy and constellation of classes, and is therefore only relatively autonomous as a political entity. Where class dominance is weak for historical and other reasons (e.g. lack of a large land-owning class), such as in some development States in East Asia (Wade, 1990), the State becomes a powerful entity.

The above broad discussion of politics at the State and city levels does not cover politics that is being played at all levels, including the level of projects. Hence, it is necessary to briefly discuss how stakeholders may be managed at the project level.

8.16 Stakeholder management

At the tactical level, project managers need to devise strategies to manage stakeholders. The first stage involves identifying all stakeholders and categorizing them using a power-interest grid (Figure 8.6).

Stakeholders with high power and economic interest in the project are key stakeholders (e.g. sponsors, contractors, and the State) and should be managed closely. Those with high power but low interest in the project (e.g. regulators) only need to be kept satisfied. Stakeholders with high interest in the project but low power (e.g. affected residents or passive equity investors) need to be informed. Finally, stakeholders with low interest and low power (e.g. general community) are peripheral to the project, and need only be monitored.

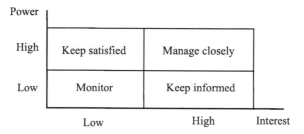

Figure 8.6 Power-interest grid.

There are many simple principles of stakeholder management. Some examples include

- taking stakeholder power and interests into account in making important decisions;
- understanding stakeholder concerns;
- listening and communicating with stakeholders at appropriate intervals;
- being objective and fair to all stakeholders; and
- quashing rumors, by responding with objective data since speculation leads to all sorts of stories.

In summary, to manage stakeholders effectively, project managers need to convey intentions, build rapport, provide information, listen attentively, understand stakeholders' perspectives, correct misconceptions, and develop genuine interest in solving problems.

Questions

1   What are the problems with conventional public procurement systems?

2   What are the advantages and disadvantages of a public-private partnership procurement system?

3   The Eppawela phosphate mining project in Sri Lanka was delayed for many years because of opposition to the scheme. In March 2000, farmers, trade unions, environmentalists, residents, and priests staged a massive street protest to block the signing between the government and a US-led consortium to mine the phosphate. The project covered an extensive area, and would involve the relocation of 12,000 people from 26 villages. Many of these villages were historic sites dating as far back as third century BC. Farmers would lose their land, and many Buddhist temples, schools, and government buildings would be destroyed. Residents also complained of wall cracks in their homes after a pilot project commenced.

   The Interior Development Minister said that the decision to go ahead with the deal was based on the enormous deposits found, and it had to go to a foreign firm because of lack of local expertise and capital. But critics said the government would only get $5 per ton, way below the market price of about $50 per ton. Environmentalists said the annual phosphate output would jump from the current 40,000 tons to 1.2 million tons, well above the 350,000 ton ecological limit.

   a) Who were the key stakeholders?
   b) How could the matter be resolved?

Risk Management Framework

9.1 Risk and uncertainty

In the previous chapter, we consider the structure of project finance in terms of the nexus of key contractual relations. This chapter builds on this base by introducing a risk management framework for projects to better understand risk allocation among parties.

Risk management has been called the "new religion" (Bernstein, 1996) with much promise as well as confusion. By "risk" we mean the probability and impact of the outcomes of a variable. Thus, understanding risk requires a clear grasp of probability theory and assessment of impacts. Probability theory is considered in the next section, and impact assessment for social projects has been dealt with in Chapter 7.

The term "uncertainty" is sometimes used to refer to outcomes that are unpredictable, that is, the probability of occurrence is unknown or not possible to estimate (Knight, 1921). If we toss a fair coin, the probability of obtaining a head may be estimated as 0.5. One may reason that there are only two outcomes, head or tail, and both are equally likely for a fair coin. Hence, the probability of obtaining a head is 0.5, and this probability is obtained by pure reasoning or a "thought experiment."

Alternatively, one can toss a coin many times and an empirical estimate of the probability of obtaining a head is the number or frequency (f) of heads in n tosses, that is, $p = f/n$. This is the frequency approach to probability, and it is based on long-run frequencies. This approach is inferior to the pure reasoning method for simple cases where it is easy to deduce probabilities and one can dispense with the need to conduct tedious experiments. However, if the coin is biased, the empirical approach is superior. If there are 300 heads in 1,000 tosses, an estimate of the probability of obtaining a head is 0.3.

Another method of obtaining probabilities is to use subjective judgment, or "gut feel." It is used when both the thought experiment method and frequency approach cannot be applied. For instance, the probability of a high vacancy rate for a hotel cannot be derived from thought experiment. If the hotel is new, there is also no data to derive historical frequencies to compute probabilities. Hence, one relies on subjective assessments. In many business situations, the use of subjective assessments is a common approach.

In contrast to the above three methods, the probability of certain events is sometimes argued to be unknowable. We simply do not have the knowledge or appropriate mathematical model to predict it. Knight (1921) argued that risk is calculable within reasonable precision while uncertainty is not calculable. Keynes (1936, pp. 149–150) echoed the same sentiment:

> The outstanding fact is the extreme precariousness of the basis of knowledge on which our estimates of prospective yield have to be made. Our knowledge of the factors which will govern the yield of an investment some years hence is usually very slight and often negligible. If we speak frankly, we have to admit that our basis of knowledge for estimating the yield ten years hence of a railway, a copper mine, a textile factory, the goodwill of a patent medicine, an Atlantic liner, a building in the City of London amounts to little and sometimes to nothing; or even five years hence. In fact, those who seriously attempt to make any such estimate are often so much in the minority that their behavior does not govern the market.

Similarly, Shackle (1979) has objected to the view that the future is knowable. This is because the future is dependent on the past and present choices we make. If there are human choices, the future is not determinate.

We will follow current practice and use the terms "risk" and "uncertainty" interchangeably. This is because the likelihood of occurrence of outcomes is, in many cases, our subjective guesses or beliefs. It may be difficult to draw a clear line between a knowable or unknowable belief. Knight, Keynes, and Shackle may have drawn the line too finely.

9.2 Probability

Since risk is the probability and impact of the outcomes of a variable, the next step in understanding project risk is to have a clear grasp of probability theory. What follows is a brief summary, and the details can be found in any standard text on probability theory.

The term "variable" is understood to be a random variable where the word "random" means the variable takes on different outcomes rather than purely haphazard or "noisy." Often, the term "random" is omitted. A variable is usually denoted by capital letters (e.g. X) and a specific outcome is denoted by x. Thus, one can write $P(X) = 0.5$, or $P(x = 2) = 0.1$. In this book, the lower case x will be used to do double work for notational convenience.

Let x be a random variable whose value is the number of dots facing upwards when a fair die is tossed. Then x takes the values and associated probability distribution given in Table 9.1. Each outcome is a sample point, and the set of all possible outcomes

$$\Omega = \{1, 2, 3, 4, 5, 6\}$$

is the sample space or universal set.

x	Probability
1	1/6
2	1/6
3	1/6
4	1/6
5	1/6
6	1/6

Table 9.1 Probability distribution of x.

An event E is a subset of the sample space. Thus,

$E_1 = \{3\}$,
$E_2 = \{2, 4, 6\}$ (i.e. die shows an even number), and
$E_3 = \{1, 3, 5\}$ (i.e. die shows an odd number)

are events.

Events A and B are mutually exclusive if whenever A occurs, then B will not occur or vice versa. Thus, E_1 and E_2 are mutually exclusive, but E_1 and E_3 are not mutually exclusive.

The probabilities must satisfy certain axioms or conditions, namely,

- $P(\Omega) = 1$;
- $0 \le P(E) \le 1$ for every event E; and
- $P(E_1 \cup E_2 \cup E_3 \ldots) = P(E_1) + P(E_2) + P(E_3) + \cdots$ if the events are mutually exclusive.

The last axiom implies that the probability of any of one of the events occurring is the sum of probabilities of mutually exclusive events. It can be proved using a Venn diagram (Figure 9.1). The universal set is drawn as a rectangle, and each event is shown as a circle. If A and B are mutually exclusive events, the circles do not overlap. Thus, the probability of A or B occurring is the sum of probabilities. This reasoning may be extended to more than two events.

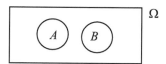

Figure 9.1 Venn diagram for mutually exclusive events.

The probability of the joint occurrence of two events, or the probability of A and B occurring, is denoted by $P(A \cap B)$. If A and B are not mutually exclusive,

$$P(A \cup B) = P(A) + P(B) - P(A \cap B).$$

Again, this result may be proved using a Venn diagram (Figure 9.2). Since the events are not mutually exclusive, their joint occurrence may be represented as an overlap between the two circles. Hence, the probability of event A or B occurring is the sum of probabilities less the overlap that has been counted twice.

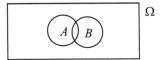

Figure 9.2 Venn diagram for non-mutually exclusive events.

Example

For the data in Table 9.1, let

$E_1 = \{3\},$
$E_2 = \{2, 4, 6\},$ and
$E_3 = \{1, 3, 5\}.$

Then

$P(E_1) = 1/6$
$P(E_2) = 3/6$
$P(E_3) = 3/6$
$P(E_1 \cup E_2) = P(E_1) + P(E_2) = 4/6$
$P(E_1 \cup E_3) = P(E_1) + P(E_3) - P(E_1 \cap E_3) = 1/6 + 3/6 - 1/6 = 3/6.$

Sometimes, the sample space is reduced and the conditional probability of event A given B is

$P(A| B) = P(A \cap B)/P(B).$

Using the same example,

$P(E_1| E_3) = P(E_1 \cap E_3)/P(E_3) = (1/6)/(3/6) = 1/3.$

The reduced sample space is $\{1, 3, 5\}$, and the probability of getting a $\{3\}$ is 1/3 since it is one of three possible odd outcomes.

Two events are said to be independent if

$P(A \cap B) = P(A)P(B).$

For example, since

$P(E_1 \cap E_3) = 1/6$, and
$P(E_1)P(E_3) = (1/6)(3/6) = 1/12$,

the events are not independent. Events are independent if the occurrence of one event does not affect the probability of occurrence of the other. For instance, the probability of having two sixes in two consecutive tosses of a fair die is

$P(6 \cap 6) = P(6)P(6) = (1/6)\ (1/6) = 1/36.$

The occurrence of a six in the first toss does not affect the probability of obtaining a six in the second toss.

9.3 Discrete and continuous variables

The die experiment discussed above is commonly found in textbooks because it provides a simple way of conveying the basic concepts of probability. To apply these concepts in practice, we often need to move away from cards and die examples.

A variable does not need to take discrete or integer values. If y is a variable whose value is the length of a stick, then y is a continuous variable that takes on real values (e.g. 1.201m).

For continuous variables, $P(y = 1.201)$ is theoretically zero. This problem is the same as spinning a needle on a disk; the probability that the needle points to a particular value is zero. Hence, we cannot use a table such as Table 9.1 to show its probability distribution. What we know when spinning the needle is the probability that it lies within a particular wedge when it is at rest. Hence, instead of a probability distribution, we have what is called a probability density function such as the one shown in Figure 9.3. It is denoted as $f(y)$. Again, it can be seen from the figure that $P(y = a)$ or $P(y = b)$ is theoretically zero. However, we are able to find $P(a \le y \le b)$, the probability that y lies between a and b. This probability is given by the shaded area. Since one of the axioms of probability is that the probabilities must sum to one, the total area under the density curve is one.

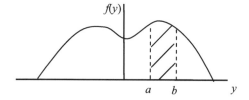

Figure 9.3 Probability density function.

Mathematically, we write

$$P(a \leq y \leq b) = \int_a^b f(y)\,dy.$$

In practice, it will be difficult to evaluate the integral analytically for the irregular density function in Figure 9.3. Hence, simpler curves are often used for theoretical and practical reasons. A commonly used density function is the bell-shaped normal distribution curve

$$f(y) = \frac{1}{\sigma\sqrt{2\pi}} \int e^{-\frac{1}{2}\left(\frac{y-\mu}{\sigma}\right)^2} dy \qquad (9.1)$$

where μ is the mean of the distribution and σ is the standard deviation, the two parameters that define a normal distribution function. Another possibility is the uniform density function

$$
\begin{aligned}
f(y) &= h && c \leq y \leq d \\
&= 0 && \text{otherwise}
\end{aligned}
\qquad (9.2)
$$

This function is shown in Figure 9.4. Since the area of the rectangle must sum to one, $h = 1/(d-c)$.

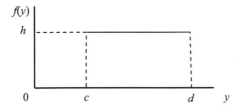

Figure 9.4 Uniform density function.

9.4 Moments

The mean and variance summarize or tell us something about the location and shape of a density function. They are special cases of what we call moments. These moments are used to tell us something about a random variable and its probability distribution.

Define the rth moment about the origin (or simply the rth moment) as

$$E[x^r] = \int x^r f(x)\,dx. \qquad (9.3)$$

The left hand side of the equation is merely notational, and E[.] is called the expectation operator. It is understood that the limits of integration covers the range of values or domain of x. Hence, the first moment about the origin is

$$E[x] = \int xf(x)\, dx. \tag{9.4}$$

It is more commonly called the mean and is denoted by μ. The second moment about the origin is

$$E[x^2] = \int x^2 f(x)\, dx, \tag{9.5}$$

and so on.

The rth moment about the mean (or rth central moment) is defined as

$$E[x - \mu]^r = \int (x - \mu)^r f(x)\, dx. \tag{9.6}$$

Hence, the second central moment is

$$E[x - \mu]^2 = \int (x - \mu)^2 f(x)\, dx. \tag{9.7}$$

It is also called the variance, and is denoted by σ^2 and σ is the standard deviation. If x is a discrete variable, the integral in the above equations is replaced with the summation sign.

As noted earlier, the first two moments, the mean and variance, are commonly used in practical applications. Higher moments are used in cases where the distribution is not symmetrical (i.e. skewed), and the mean and standard deviation no longer adequately describe such a distribution. It may be noted in passing that returns from stock prices may not be symmetrically distributed about the mean or follow a normal distribution. The "tails" tend to be fatter.

Example

Find the mean and variance of the uniform distribution in Figure 9.4 if $c = 2$ and $d = 6$.

Here, $h = \frac{1}{4}$ so that $f(x) = \frac{1}{4}$ and the mean is given by

$$\mu = \int_2^6 xf(x)\, dx = \int_2^6 (x/4)\, dx = 4.$$

The variance is given by

$$\sigma^2 = \int_2^6 (x-4)^2 f(x)\, dx = \int_2^6 ((x-4)^2 / 4)\, dx = 4/3.$$

Example

Use Table 9.2 to find the mean and variance of the discrete variable x.

x_i (%)	Probability, p_i
2	0.1
3	0.3
4	0.5
5	0.1

Table 9.2 Probability distribution of x.

Since x is a discrete variable, we replace the integral in the moment equations with a summation sign. Hence,

$$\mu = \Sigma \, x_i p_i = 2(0.1) + 3(0.3) + 4(0.5) + 5(0.1) = 3.6.$$

$$\sigma^2 = \Sigma \, (x_i - 3.6)^2 p_i$$
$$= (2 - 3.6)^2 0.1 + (3 - 3.6)^2 0.3 + (4 - 3.6)^2 0.5 + (5 - 3.6)^2 0.1 = 0.64.$$

It can be seen from these examples that finding the moments and central moments of simple density functions such as the uniform density function is straightforward. However, some density functions such as the normal distribution curve in Equation (9.1) can be difficult to integrate. A solution is to do solve the integral numerically and present the results in the form of a statistical table found at the back of statistical textbooks.

9.5 Risk exposure

It has been noted that risk is the probability and impact of the outcomes of a variable.

From this definition of risk, it seems reasonable to define

Risk exposure = Probability × Impact

i.e.

$$E_i = P_i \times I_i. \tag{9.8}$$

For instance, if the probability of fire is 0.001 and the expected damage is $1m, its risk exposure is $1,000.

It follows from the definition that events with high probabilities and impacts present significant risk exposures (Figure 9.5).

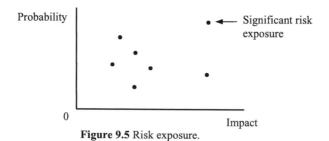

Figure 9.5 Risk exposure.

Sometimes, the frequency of occurrence of an event (F_i) is added to Equation (9.8) so that

$$E_i = P_i \times I_i \times F_i. \tag{9.9}$$

As noted at the beginning of this chapter, probability may have a frequency definition so that F_i is already partly subsumed under P_i. Further, F_i is also partly subsumed under I_i since the frequency of occurrence of an event will affect the impact. For these reasons, Equation (9.9) will not be used to avoid conflating the issues.

This short introduction completes our discussion on the concept of risk exposure and the underlying probability theory. The next step is to consider the scope and other aspects of risk management.

9.6 Scope of risk management

The scope of risk management consists of four main elements (Waring and Glendon, 1998), namely,

- objects of risk management;
- risk management contexts;
- objectives of risk management; and
- methods of risk management.

These elements are discussed below.

9.7 Objects of risk management

The objects of risk management are the hazards or threats that result in undesirable outcomes. The main objects concern

- financial loss or gain from price or output changes;
- opportunistic behavior and other contractual problems;

- political or social problems;
- physical damage (e.g. fire);
- technical problems (e.g. soil conditions);
- safety and health;
- regulatory problems;
- tax changes; and
- environmental damage.

These objects of risk management are generally easy to identify in a project. However, there are objects that are unknowable to the individual, project team, or organization, and when such an incident happens, it is viewed as unexpected or a "freak accident." Consequently, proper identification of the objects of risk management should be a rigorous exercise and not just a routine procedure. A checklist should be used. Some of these objects of risk management are briefly discussed below.

The pervasive influence of the market on prices and output has led to many speculations about market stability or instability as well as the role of expectations in stabilizing or destabilizing markets. For classical and neoclassical economists, markets are generally stable (i.e. tend towards equilibrium) and depressions are viewed as temporary or slight deviations before prices, wages, rents, and interest rates adjust "relatively quickly" or "continuously" to bring the goods, labor, land, and money markets into equilibrium at full employment level. In essence, demand and supply shifts due to external shocks (such as the weather, technology, or discovery of gold) lead to fluctuations in output and prices. These changes in prices and outputs are known or signaled quickly to buyers and sellers and they adjust their behavior accordingly.

As is well known, Keynes (1936) argued that, on the contrary, wages are "sticky" because of long-term labor contracts (and trade union power in some cases) and, consequently, the labor market cannot adjust rapidly by bringing down wages in the short turn. Further, even if wages can be adjusted downwards in a depression, it will adversely affect consumer demand. The fall in demand will, in turn, affect output and employment.

If wages (i.e. prices of labor) cannot adjust quickly, then excess demand can only be cleared through quantity rather than price adjustments, resulting in high unemployment. In Figure 9.6, D is the demand curve, S is the supply curve, P is unit price, and Q stands for quantity demanded. P^* is the sticky or relatively fixed price of a good. It is above the equilibrium level that brings demand and supply into balance. At price P^*, the quantity supplied is b, and quantity demanded is a, and the excess output is therefore $b - a$.

If price does not adjust downwards rapidly to remove or clear the excess output and bring demand and supply into equilibrium, quantity adjustments as shown by the arrow will take effect. Faced with surplus goods, firms can either lower the price or cut down production. Aggressive marketing may help individual firms, but it is unlikely to raise overall demand substantially in a depression.

Generally, firms prefer to maintain the output price to avoid hefty losses and cut costs. However, if the nominal price of labor (w) is sticky and wages are a major portion of unit cost, then the real wage w/p where p is the general price level actually rises because of falling prices in a recession. If firms are unwilling to lower price and wages cannot be reduced substantially, they will have to cut production, which often entails firing workers. Hence, unemployment rises.

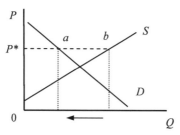

Figure 9.6 Quantity adjustment.

As unemployment rises, workers' incomes fall, and the overall demand for goods and services will shrink. Keynes argued that, in the face of gloomy prospects, the "animal spirits" of businessmen take over in a recession, and firms are unwilling to undertake investment if the outlook is bleak. Hence, Keynes supported increasing State expenditure or "pump priming" to lift the economy out of depression. In a depression, prices are depressed, so the economy can tolerate a little inflation caused by the increase in government expenditure that is financed by printing money (or, more politically correct, by increasing money supply). Keynes argued against raising taxes for political reasons, and also because taxing the private sector to pay for public spending merely transfers buying power and does not raise overall demand. As the economy recovers, tax revenues rise sufficiently to cover the government deficit.

That, in brief, is the theory. In practice, governments are pressured to increase spending, resulting in the high inflation of the 1970s. Consequently, Keynesian "pump priming" became unpopular, and the main objective of macroeconomic policies in many of the developed countries shifted from ensuring full employment to booting out inflation. This entailed sharp reductions in government expenditure, particularly in welfare payments.

Although a new breed of "new classical" economists still assume that markets clear continuously, the current consensus since the financial crises of the 1980s and 1990s is that markets are inherently unstable and require supporting institutions to function properly (World Bank, 2002).

There are two other very different views on market instability that should be highlighted. Unlike the equilibrium economists discussed above, the Austrian economists (Kirzner, 1985) view the market more as a discovery rather than as an adjustment or equilibrating process. In this perspective, the entrepreneur continuously looks for errors in valuation (i.e. discovery of bargain prices) and business opportunities that have been overlooked. This means the market is not in a stable equilibrium.

Finally, Schumpeter (1942) argued that the market is also a creative process. Entrepreneurs continually create new products, services, transport means, forms of financing, sources of materials, communication channels, markets, and organizing principles that destroy the old ones and even the established firms. The "gale" of innovation or creative destruction creates and destroys. This implies that markets are inherently unstable.

Opportunistic behavior and other contractual problems is the second object of risk management. The presence of opportunistic behavior puts into sharp focus the important role of human rather than technical factors in managing risk. A firm enters into a contract with a potential worker, purchaser, or supplier because it wants to

- share or shift the price, output, and other risks;
- provide incentives for producing quality goods; and
- prevent hold-up.

For instance, a contractor may enter into a long-term contract with a steel supplier to fix the price of steel now through mutual agreement for steel to be delivered some time in future. Both the contractor and supplier share the risk of future price fluctuations.

In addition, a contract may provide incentives for the supplier to deliver the correct quality of steel. Otherwise, the supplier may deliver inferior steel. Hence, proper requirements, specifications, measurement, and enforcement are required for contracts to function properly.

Contracts are also used to prevent opportunistic behavior or "self-interest seeking with guile" (Williamson, 1975). It arises because contracts are necessarily incomplete as a result of

- cognitive limits or bounded rationality (Simon, 1957), resulting in the inability to foresee all future contingencies or uncertainties; and
- linguistic imprecision in writing contracts.

In addition, there is the possibility of a hold-up because of asset specificity (Williamson, 1975). It refers to the extent to which assets can be redeployed. For instance, if party A wishes to buy a particular good from B without a contract and, after learning that B has invested in some specialized equipment, A may sought to renegotiate the price (i.e. indulge in opportunistic behavior) knowing that B's bargaining position is now weaker. B's investment has been "locked in" and she faces a "hold-up" problem.

There are many ways in which one party can be tied, for example through

- specialized equipment;
- specialized labor;
- specific site; and
- specific time or period.

It is easy to think of situations where equipment, workers, and sites cannot be readily redeployed. Temporal specificity refers to a situation where, for example, goods may be perishable and the supplier's bargaining position weakens as time goes by.

9.8 Risk management contexts

The risk management contexts refer to the inner and outer contexts that set the scene for managing project risk. The inner context concerns how risk is managed within the organization. The outer context refers to the organization's environment. These contexts are discussed below.

Inner context

The inner context may be divided into three levels of analysis, namely,

- the individual;
- groups; and
- the organization as a whole.

At the individual level, one could consider a person's motivation to manage risk, appetite for risk or degree of risk aversion (see Chapter 10), limited cognition (i.e. bounded rationality), and level of skills in managing risk.

A person's motivation to manage risk depends on many factors such as pay, skill level, position (responsibility), and apathy. Many people do not work well when underpaid, and some do not put in sufficient effort even when overpaid. The lack of interest or care can spiral into a serious problem resulting in costly rework or accidents. If a worker's skill level is low, training can be provided. Fixing general apathy among workers is a much more difficult problem for management to tackle.

It is well known that individual cognition and decision-making are known to contain biases such as

- stereotyping;
- limited search for alternatives;
- deciding too quickly;
- generalizing from insufficient cases;
- tendency to use readily available information;
- tendency to ignore fundamentals and give excessive weight to recent information;
- inadequate methodology;
- viewing positive outcomes as more probable than negative outcomes;
- downplaying the risks;
- following the leader for reasons such as power relations, laziness, insufficient information, insufficient preparation or incompetence; and
- refusing to believe a run of bad luck or long runs in sequence of random tosses of a fair coin (Tversky and Kahneman, 1974).

At the group level, team decision-making is affected by

- power relations;
- divergent goals;
- tendency to rely on others to do the homework, talking or decision-making;
- group thinking or desire for conformity than the right decision; and
- tendency to make risky decisions because of diffusion of responsibility or the chance to get back at an opponent.

In the case of power structures, the separation of corporate ownership and control (Berle and Means, 1932) raises the question of what exactly do managers optimize or satisfice (Simon, 1957), or do they even optimize or satisfice at all. "Satisficing" refers to a limited search for a satisfactory solution ("that will do") rather than an exhaustive search for the optimum point. Such behavior has generally been downplayed in the neoclassical economics literature by simply arguing that managers act *as if* they optimize (Machlup, 1946).

The standard answer to the question what managers actually do is that managers maximize long-run profit or shareholder wealth. There is little support for the alternative view that managers maximize sales or growth (Marris, 1964; Baumol, 1959). If managers optimize their remuneration or status, then it is conditional on profits or share price. From a radical perspective, Marglin (1976) has argued that managers are more interested in control rather than productivity or efficiency. If this claim is true, then risk management systems are primarily control systems.

At the organization level, how risk is managed depends on its

- corporate culture or appetite for risk;
- procedure and time for decision-making;
- corporate strategy, such as the desire to enter a new market; and
- resources to undertake the risk.

Outer context

The outer context concerns the organization's environment, that is, the political, economic, social, legal, and technological factors. These factors will affect how a firm manages its risks. For instance, firms in cyclical and volatile industries will need a different set of strategies and tools from organizations operating in more stable environments.

9.9 Objectives of risk management

The objectives of risk management are to

- eliminate, reduce or control pure risks; and
- benefit or hedge from speculative risks.

Risks that have only downside, such as fire damage, are pure risks. Speculative risks such as currency fluctuations and changing market conditions have both downside losses and upside profit opportunities. A risk management strategy should hedge the organization from downside risk and benefit from upside opportunities.

Sometimes, the objectives of risk management are called risk resolution strategies.

9.10 Methods of risk management

Finally, the methods of risk management consist of a risk management system (RMS) for managing risk within an *organization* and a risk management process (RMP) within a *project*.

An RMS includes

- the strategy, policy and objectives;
- organizing, resourcing, and planning;
- implementing plans;
- review and monitoring; and
- feedback.

Such a system is often subjected to external audits or certification to comply with certain local or international standards. In addition, the RMS should be integrated with other management systems.

Generally, an RMS encompasses a number of risk management principles such as those shown in Table 9.3. These principles are adapted from the Software Engineering Institute (2006).

Principle	Elements
Global view	View project development within context of higher level definition, design and development and recognize potential opportunities and adverse effects
Forward looking	Anticipate potential outcomes and manage them
Open communication	Encourage free flow of information, enable formal and informal communication, and value the individual voice
Integration	Integrate management systems and make risk management an integral part of project management
Continuous process	Manage risk continuously over the project cycle
Shared product	Common purpose, vision, and ownership with a focus on results
Teamwork	Cooperation and polling of resources and talents

Table 9.3 Principles of risk management.

Unlike the risk management system, the RMP operates at the project level as a sequence of steps taken to achieve the objectives of risk management. It begins as part of project planning and ends with the project close-out. The steps in a risk management process include

- risk identification;
- risk assessment;
- risk prioritization;
- risk strategies and contingency planning; and
- risk monitoring and review.

These steps are outlined below.

Risk identification

The first step in the risk management process is to identify the types of project risks. These risks may be classified by broad categories, project phases, or a combination of both methods (see Table 9.4).

The identification of risks is an ongoing process as circumstances change. Such risks may be identified at regular risk management meetings.

Type of risk	Items
Completion	• Time (e.g. delays due to weather, strikes, or late delivery of equipment) • Cost, theft, fire • Quality • Design, technical • Land title
Counter-party	• Inability of other party to pay or perform
Political	• Expropriation, nationalization • Restrictions on repatriation of funds • Lack of commitment, change of government or ideology • Trade embargo • Change in tax regime
Force majeure	• Natural disasters • War and civil unrest
Financial	• Interest rate and inflation • Currency
Input	• Price, quantity, quality
Market	• Demand
Insurance	• Adequate cover
Environmental	• Pollution • Destruction
Operation	• Labor issues, poor maintenance, supplies, rising costs, defects
Regulatory	• Uncertain and complex laws, price controls, corruption, false reporting • Permits and licenses
Residual value	• Uncertain value of asset at the end of concessionary period
Technological	• Sub-optimal facility on completion due to technological changes

Table 9.4 Risk identification.

Risk assessment

In the risk assessment phase, probabilities and impacts are assigned to each risk. That is, since

Risk exposure = Probability × Impact

or

$$E_i = P_i \times I_i,$$

the task in risk assessment is to determine P_i and I_i for each risk. It is more important to avoid oversight rather than to accurately identify probabilities and impacts. For instance, the probability of war and its impact are difficult to assess accurately so one has to rely on rough numbers or simple rating scores to estimate the risk exposure.

Risk prioritization

In this step, the risks are prioritized using risk exposure. Items that have high exposures warrant close attention.

Plots similar to Figure 9.5 may be used. In many cases, difficulties in assigning probabilities and impacts lead to the splitting of the axes into three simple categories (Low, Medium, and High), resulting in a risk matrix (Figure 9.7). Particular attention is paid to areas marked "X", that is, risks with high probabilities or impacts.

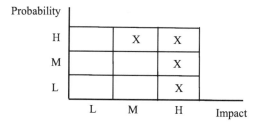

Figure 9.7 Risk matrix.

Risk strategies and contingency planning

After identifying, assessing, and prioritizing risks that warrant attention, the next step is to consider strategies to avoid, mitigate, transfer, investigate, share or accept these risks.

The strategy to use depends on the risk exposure and methods available (Table 9.5). Risk strategies are also called risk responses.

Type of risk	Risk strategies
Completion	• Project management • Bonds, insurance, warranties • Procurement contracts, selection of contractors • Land surveys, due diligence
Counter-party	• Bonds, contracts, guarantees, escrow accounts
Political	• Due diligence, political risk insurance
Force majeure	• Insurance
Financial	• Derivatives, price indexes, reserves
Input	• Contracts • Diversification • Buffer
Market	• Feasibility study • Off-take contracts • Options
Environmental	• Assessment, prevention, control measures
Operation	• Appropriate labor policies, selection, training, • Procedures • Operations and maintenance contracts
Regulatory	• Due diligence
Residual value	• Forecasts, contracts
Technological	• Due diligence

Table 9.5 Risk strategies.

Risk monitoring and review

The final step in the risk management framework is to monitor the risk management process continuously and conduct periodic reviews on the effectiveness of the risk management system and process in resolving risks. It includes

- creating a central depository for risk information and documentation;
- assigning of risk responsibilities;
- creating a risk summary report; and
- regular progress meetings and reports.

Risks do not stay the same. Circumstances, rules, priorities, and people change. Hence, risks will change over time, and require constant monitoring and periodic reviews. A major incident will trigger an immediate review of the risk management system and process.

Questions

1   If a fair coin is tossed repeatedly until the first head appears, determine the sample space.

Note: Theoretically, the sample space in this example is countable but infinite. Hence, the axioms of probability need to be modified slightly to take into consideration this possibility.

2   If each number in a countable set $\{x_1,..., x_n\}$ has equal weight or equal chance of occurring, prove that

a) $E[x] = \mu = (\Sigma x_i)/n$;

b) $Var(x) = (1/n)\Sigma(x_i - \mu)^2$

3   If a card is selected from a deck, compute the following probabilities:
a) $P(Ace)$;   [4/52]
b) $P(Ace\ of\ spade)$;   [1/52]
c) $P(King\ or\ Ace)$; and   [8/52]
d) $P(King\,|\,Face)$.   [4/16]

4   If the probability of a crane breaking down on a given day is 0.02, what is the probability that both cranes will break down in the same day? [0.0004]

5   If the impact from the breakdown of both cranes will result in a loss of $30,000 per day, what is the risk exposure if the cranes need two days to replace? [$24]

6   What are the limitations of a risk management framework?

7   Who is responsible for implementing a risk management framework in
a) An enterprise?
b) A project?

8   You have searched a website on possible suppliers for tiles. What are your risks and exposure?

9   Explain why many mega-projects tend to face significant cost over-runs from a risk perspective.

10

Risk, Insurance, and Bonds

10.1 Insurable and uninsurable risks

We saw in Chapter 9 that insurance plays a major role in risk management by shifting risk from a risk-averse party to another party that is better able to handle it by paying a premium. The basis of the comparative advantage in risk bearing lies in risk pooling based on the law of large numbers. Obviously, if the risk pool is small, the comparative advantage of the insurer to bear the risk is eroded.

If the exchange is freely carried out and both parties are better off, the transaction is a Pareto improvement.

Many project risks such as fire damage, negligence, and theft are insurable. Probabilities and impacts may be computed to derive fair premiums. However, as pointed out by Borch (1990) and many others, the concept of insurability need not rely on the ability to compute probabilities and impacts. As long as both parties freely agree to an insurance contract, the risk is insurable.

Some risks are uninsurable in the sense that insurance markets for these risks do not exist. There are several reasons that may make some risks uninsurable (Arrow, 1963):

- adverse selection;
- moral hazard;
- transaction costs; and
- nondiversifiable risks.

In adverse selection, insurers are unable to accurately differentiate high risk from low risk groups and therefore tend to charge a common premium. This premium is too high for the low risk group, and many members from this group decide not to insure. Since the pool now consists primarily of high risk buyers of insurance (i.e. an adverse selection has occurred), there will be more claims. The premium will rise if the insurer is to stay in business. Eventually, the risk may be uninsurable because of excessive claims or high premiums. Adverse selection is also called the "hidden information" problem because buyers of insurance know their risk profiles better than insurers.

Moral hazard is a "hidden action" problem where the insured can cheat on his obligations such as taking proper care to avoid fire, theft, or car accidents instead of carelessly reasoning that "it is insured anyway." If the insured is careless and not penalized for it, then insurance claims or premiums will rise and the risk may become uninsurable.

In both adverse selection and moral hazard, the insurer has another option. This is to search for the hidden information or supervise the hidden action. However, both approaches incur transaction costs, and if these costs are too high, the risks are again uninsurable.

Social risks such as war, social unrest, and natural disasters are largely nondiversifiable and this may impede the development of insurance on them. This is the reason why political risk insurance against civil unrest, expropriation of assets, currency convertibility, contract repudiation, and license cancellation takes so long to develop in project finance, and even today, there are few insurers who are willing to underwrite this risk.

10.2 Mechanisms to create insurance markets

Many mechanisms have been devised to alleviate the problems of adverse selection. These include

- screening of applicants, such as health and income screening, experience rating, background checks, and so on;
- blacklisting undesirable buyers, which is also a form of screening;
- limits on the maximum amount insured;
- compulsory insurance to prevent low-risk groups from opting out; and
- menu of polices each with different premiums and coverage to cater to low-risk and high-risk groups.

In the case of moral hazard, the mechanisms to create insurance markets include

- coinsurance or co-sharing arrangements;
- charging higher premiums, such as in car insurance for drivers who have poor driving records;
- no-claim bonus to encourage proper care;
- inspection of premises, such as to ensure you take precautions against theft;
- exclusions; and
- deductible or excess, that is, the amount below which shall be borne by the insured party.

Finally, for nondiversifiable or partially diversifiable risks such as the impact of natural disasters, the mechanisms include

- reinsurance by insurers to diversify further, presumably because well-diversified insurers are better able to handle the risks;
- compulsory public provision of insurance;
- risk securitization, such as through catastrophe-linked bonds or "cat bonds" that pay high coupons but if the issuer (i.e. insurer) suffers a loss in

the event of a pre-defined catastrophe, the obligation to pay coupon and/or principal is deferred or forgiven; and

- risk mutualization (Borch, 1962) or risk-sharing of potential losses among those insured, between those insured and the insurer, and among insurers and investors. For instance, the payout may depend on the total value of claims by all those affected by the disaster.

The use of such mechanisms raises many questions beyond the scope of this book. Only a few issues are discussed below, and they include

- the efficiency of the insurance industry;
- degree of risk aversion;
- premium level and optimal level of coverage;
- interactions among different sources of risks; and
- self-insurance and self-protection.

10.3 Structure of insurance markets

The structure of an industry refers to the following:

- number of sellers and buyers;
- concentration, i.e. control of market share by the few largest firms;
- entry barriers; and
- product differentiation.

Early work on the US market (Joskow, 1973) showed that the property and liability insurance (PLI) industry has 1,206 firms, which is a large number. Concentration is relatively low, with the four largest firms having less than 20 per cent of market share. Despite subsequent restructuring of the industry through mergers and acquisitions and vast improvements in information and communications technologies, the industry is still competitive (Hanweck and Hogan, 1996).

One reason is that entry barriers are low, allowing a large number of firms to enter and exit the industry. The failure of natural monopolies to emerge is due to constant economies of scale. In general, small and medium size PLI firms tend to enjoy economies of scale, but large firms tend to be exhibit diseconomies (Hanweck and Hogan, 1996; Cummins and Weiss, 1993). Cummins and Van Derhei (1979) suggested that there are increasing returns to scale as the risk declines with increasing number of exposures provided the outcomes are uncorrelated across different risk categories. However, the decline in risk falls at a decreasing rate. Hence, for most part, the insurance industry operates close to constant returns to scale with low levels of concentration. There is no evidence that large firms offer lower premiums than smaller ones for similar products. Product differentiation is also limited in insurance. It is relatively easy to incorporate new clauses in a policy.

Underwriting (selling) cost has long been known to be high in property and liability insurance and for which insurers have been criticized. Joskow (1973) estimated that underwriting cost could be as high as 36 per cent of premiums for listed insurance firms. Insurance is normally bought through brokers who are either exclusive to an insurance firm or act independently. The latter is generally more efficient because of the opportunity to compare policies. However, independence may only be in theory rather than fact. The quality of broking services can vary considerably and the commission often ranges between 10–20 per cent, which is relatively high.

10.4 Degree of risk aversion

The pre-modern theory of risk is based on the expected monetary value (EMV). Suppose a fair coin is tossed and the payout is $1 if the outcome is a head and $0 if it is tail. Then

$$\text{EMV} = 0.5(1) + 0.5(0) = \$0.50. \tag{10.1}$$

If each bet is $0.50, that is, equal to the expected value, the game is said to be fair. More generally, if there are $x_1,\ldots, x_n$ possible outcomes with probabilities $p_1,\ldots, p_n$ respectively, then

$$\text{EMV} = \Sigma\, p_i x_i. \tag{10.2}$$

Suppose the payout is raised to $1,000 if the outcome is a head and $0 if it is tail. Then

$$\text{EMV} = 0.5(1000) + 0.5(0) = \$500. \tag{10.3}$$

If each bet is $500 and equal to the expected value, the game is still fair. However, not many people will play this game with a higher stake.

In 1738, Daniel Bernoulli tried to explain this phenomenon by arguing that the marginal utility of money falls as a function of wealth. This means that each additional dollar yields lower and lower utility. An additional dollar is worth less to a rich person than to a poor person.

This idea of diminishing marginal utility of money was later formalized by von Neumann and Morgenstern (1944). In Figure 10.1, w denotes the level of wealth of an individual, and $U(w)$ is the associated utility level. The utility function may be normalized by assigning $U(\$0) = 0$, and $U(\$1,000) = 1$. The utility curve is drawn concave to the origin, implying the marginal utility of money falls as the level of wealth rises.

Now, the expected utility of playing the game is

$$U_g = 0.5U(1000) + 0.5U(0) = 0.5(1) + 0 = 0.5. \tag{10.4}$$

Since $U_g < U(\$500)$, the utility of a sure sum of $500, such a person is said to be risk-averse (i.e. dislike risk) because he would not play a fair game with high stakes. The assumption of diminishing marginal utility of money has solved the paradox.

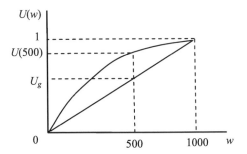

Figure 10.1 Expected utility theory.

An investor is said to be risk neutral if only the expected project returns matter, that is, the variance of the return (or risk) is neglected. Being indifferent to risk, or risk neutrality, is not a plausible assumption. Hence, Equation (10.4), which only considers expected returns or utilities, must be further modified to take into account risk.

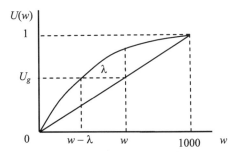

Figure 10.2 Risk premium in the general case.

Since a risk averse individual prefers the sure sum of $500 than play a $0 or $1,000 game with equal probability, he may be willing to pay a small price (λ), the risk premium (Figure 10.2), to avoid playing the game and still achieve the utility level given by U_g. Hence,

$$U_g = U(w - \lambda). \tag{10.5}$$

Since U_g is the expected utility from the gamble, it may be written as

$$U_g = E[U(w + z)] \tag{10.6}$$

where $E[z] = 0$ and $var(z) = \sigma^2$. Note the normality of z is not required. Combining Equations (10.5) and (10.6),

$$U(w - \lambda) = E[U(w + z)].\tag{10.7}$$

Both sides of Equation (10.7) may be expanded using the Taylor series

$$f(x+h) = f(x) + f'(x)h + \frac{f''(x)}{2}h^2 + \cdots\tag{10.8}$$

where $f'(x) = df/dx$ and $f''(x) = d^2f(x)/dx^2$. If we apply Equation (10.8) to both sides of Equation (10.7) and neglect higher order terms,

$$U(w-\lambda) = U(w) - U'(w)\lambda + \frac{U''(w)}{2}\lambda^2;$$

and

$$E[U(w+z)] = E[U(w) + U'(w)z + \frac{U''(w)}{2}z^2].$$

Equating both sides gives

$$U(w) - U'(w)\lambda + \frac{U''(w)}{2}\lambda^2 = E[U(w) + U'(w)z + \frac{U''(w)}{2}z^2].$$

Since $E[z] = 0$ and $E[z^2] = \sigma^2$, the above expression reduces to

$$-U'(w)\lambda + \frac{U''(w)}{2}\lambda^2 = E[U'(w)z] + E[\frac{U''(w)}{2}z^2] = \frac{U''(w)}{2}\sigma^2.$$

Neglecting λ^2 gives the Arrow-Pratt risk premium (Pratt, 1964; Arrow, 1963)

$$\lambda = \frac{\sigma^2}{2}A(w)\tag{10.9}$$

where
$$A(w) = -U''(w)/U'(w)$$

is called the degree of absolute risk aversion of the agent. Hence, the risk premium depends on

- the variance of z (i.e. σ^2), and
- the degree of absolute risk aversion (curvature of the utility curve) of the agent evaluated at w.

Equation (10.9) provides a basis for using the variance or standard deviation as a measure of risk. This means that all we need to consider when making decisions involving risk is the mean and variance of the outcomes. This approach is called the mean-variance framework.

Unfortunately, the approximation in Equation (10.9) holds only when risks are small. There are two approximations, namely,

- Taylor series linearization, implying that the curvature is not too sharp in the neighbourhood of w; and
- λ^2 has been neglected, that is, the risk premium is small, which implies that risks are small.

Further, the result does not hold in situations where the outcomes are not symmetrically distributed since it has been assumed that $z \sim (0, \sigma^2)$. For instance, the downside to a research and development project may be limited, but the upside may have tremendous potential. The variance is no longer an adequate measure of risk.

We saw in Chapter 6 that the Capital Asset pricing Model is based on the mean-variance framework. The hurdle rate is based on the weighted average cost of capital (WACC) plus a mark-up for risk (see Figure 6.3).

Apart from its applicability to only cases involving small risks, one other criticism of the mean-variance framework is that utility curves may not be concave (Friedman and Savage, 1948). It has been observed that people are not necessarily risk averse; they may be risk loving (or risk seeking) at low levels of wealth by purchasing lottery and risk averse at higher levels of wealth beyond W in Figure 10.3.

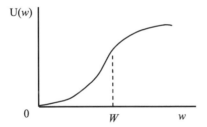

Figure 10.3 Non-concave utility curve.

A final criticism of expected utility theory is that the shape of the utility curve depends on how a problem is framed. Although Kahneman and Tversky (1979) explained it in terms of monetary values, it is possible to use a utility curve to illustrate the idea. If we take W in Figure 10.3 rather than the origin as the zero point, the curve is steeper to the left, implying that a "disutility" or loss is more painful than a gain of the same amount. In other words, the pain of losing $100 is more than the gain of $100. Consider the following games:

Game 1
Choice *A*: A 50 per cent chance of gaining $1,000
Choice *B*: A sure gain of $500

Game 2
Choice *A*: A 50 per cent chance of losing $1,000
Choice *B*: A sure loss of $500

In Game 1, the problem is framed as a gain and most people are risk averse and pick *B*, a sure gain of $500. In Game 2, the sure loss of $500 encourages many people to take a gamble and choose *A*. Hence, even though both games are numerically equivalent, people make different choices depending on how the problem is framed.

There are many other problems with expected utility theory, and the theory has both been defended and criticized (see Watt, 2002).

10.5 Premium and optimal coverage

Insurance premiums that are actually paid for a policy should not be confused with the (theoretical) risk premium one is willing to incur to avoid playing a game and still achieve a certain level of utility.

Insurance premiums are considered fair if they equal the expected monetary value after accounting for underwriting costs and normal profit for insurers. If premiums are fair, how much coverage should be purchased?

From the point of view of the insurer, it is unwise to provide coverage greater than the market or replacement value of the asset. Otherwise, the policy holder may have an incentive to torch the asset.

Let

W = initial wealth
C = amount of insurance coverage purchased
L = expected loss
p = probability of loss

There are two possible situations, and these are shown in Table 10.1 (Mossin, 1968). In the first case, nothing happens (i.e. there is no loss) with probability $1 - p$ and the insured party pays only the fair insurance premium (= pC) and her wealth position is $W - pC$. In the second case, the loss happens with probability p and the insured party losses L but makes a claim for amount C. Since the insurance premium pC still needs to be paid, her wealth position is $W - pC + C - L$.

Situation	Wealth	Probability
No loss	$W - pC$	$1 - p$
Loss	$W - pC + C - L$	p

Table 10.1 Payoffs and optimal insurance.

The expected utility is

$$m = (1 - p)U(W - pC) + pU(W - pC + C - L). \qquad (10.10)$$

To determine the amount of insurance coverage, we partially differentiate m with respect to C and equate it to zero:

$$\frac{\partial m}{\partial C} = (1 - p)(-p)\frac{\partial U(W - pC)}{\partial C} + p(-p + 1)\frac{\partial U(W - pC + C - L)}{\partial C} = 0.$$

Simplifying,

$$\frac{\partial U(W - pC)}{\partial C} = \frac{\partial U(W - pC + C - L)}{\partial C}.$$

Since the first partial derivatives are the same and utilities are only affected by wealth levels, this implies

$$W - pC = W - pC + C - L$$

or

$$C = L.$$

In other words, the optimal coverage is to fully insure for the expected loss.

If full insurance is optimal, why do we find partial insurance in practice? The reasons include

- differences in expectations on the probability of occurrence (p);
- differences in expectations on the expected loss (L);
- high premiums ($> pC$) rather than fair premiums;
- budget constraint;
- insurance is imposed by another party (e.g. lender);
- taking the gamble and under-insure;
- poor understanding of complex insurance packages; and
- interactive effects (see next section).

In the case of project finance, lenders tend to impose insurance such as political risk insurance as a condition for loan. Insurance may not be freely chosen by project sponsors.

It is difficult to disentangle the factors above as a cause of under-insurance or over-insurance. For instance, the fair premium is pC, and yet differences in assessments of probabilities or coverage will lead to different premiums independently of underwriting costs.

Further, the value of an asset may not be static such as when a facility under construction appreciates in value because of design changes or cost escalation. An escalator clause may be included in the policy so that coverage automatically rises with the value of the asset.

10.6 Interactive effects

The discussion so far considers risk in isolation. It is obvious that accidents may not happen in isolation and, if the monetary outcomes are correlated, the issues and conclusions discussed above may need to be modified. For instance, excessive rain may damage houses and fill up a dam, resulting in positive and negative outcomes. Hedging then becomes possible, and the optimal policy is not to fully insure the house against flood (Mayers and Smith, 1983).

In this sense, it is possible to view insurance as part of portfolio hedging activity.

10.7 Self-insurance and self-protection

An individual faced with a certain risk exposure need not buy full insurance coverage if he can self-insure or self-protect. Since

Risk exposure = probability of occurrence × impact,

self-insurance relates to reducing the size of impact (e.g. sprinkler system against fire) and self-protection relates to reducing the probability of occurrence of a loss, such as the installation of burglar alarms to reduce the probability of theft (Ehrlich and Becker, 1972). The interesting part about self-insurance and self-protection is that they are substitutes to market insurance and become important when insurance markets are missing or limited.

10.8 Practical considerations in insurance

The above theoretical considerations may be briefly summarized. Some risks are uninsurable and therefore fall under exclusion clauses in insurance policies. Where risks are insurable, the problems of moral hazard and adverse selection result in the use of many mechanisms to create workable insurance markets. Such markets are generally competitive because of limited economies of scale and low barriers to entry.

Although people are generally risk averse, they do take risks and under-insure, and this accounts for both concave and convex utility curves. The risk premium to avoid a risk depends on the variance of outcome and degree of absolute risk aversion (curvature). The latter depends on an individual's wealth level. Generally, the optimal coverage is to fully insure but there are many reasons for partial insurance. There are also interactive effects among outcomes that provide a hedge to only partially insure. Finally, self-insurance and self-protection are incomplete substitutes to market insurance.

In practice, the following insurance polices are usually found in building and infrastructure projects:

- third-party insurance against injury or property damage to third parties, and this generally covers the interests of all parties in the project, that is, owner, contractor, subcontractors, suppliers, and designers;
- builder's risk insurance against first-party injury and damage to property such as foundation, superstructure, temporary works, vehicles, equipment, and materials onsite, offsite, and in transit;
- worker's compensation insurance against injury to workers including medical costs, disability benefit, and death benefit;
- professional liability insurance taken by design professionals separately to cover liability out of negligence, error or omission; and
- special insurance for items such as marine structures not covered by the above.

Sometimes, these policies are combined, such as an "All Risks" insurance policy combining third-party insurance and builder's risk insurance.

There are many variations in terms of premium and coverage. For instance, the "All Risks" insurance policy is usually taken for a specific project although it can be done on annual basis to cover most of a contractor's work, particularly if the employees of a contractor work on different projects. Similarly, a plant that is used for different project sites may be covered separately for damage from use, repairs, and transport under an annual policy. If a plant is hired, the leasing company would have insured it against damage but it may claim for loss of income against a contractor for lack of adequate care in using the equipment. If a road or pipeline project is divided into sections, it may be better for the client rather than individual contractors to undertake the insurance to avoid complex claims arising from boundary problems.

If it is specific to a project, the period of cover is determined by the construction period and usually extends to 14 days after the issue of Certificate of Completion rather than a fixed date to cover possible project delays. Insurers require the contractor to

- inform them of extensions of time and changes in risks (e.g. major design changes) for which insurers are not liable;
- take reasonable precautions to avoid the damage or loss such as ensuring proper training and maintenance;
- use arbitration rather than litigation in settling disputes over an adjuster's recommendation; and
- pay additional premiums if an amount insured is to be reinstated to the original level following a claim.

An "All Risks" insurance policy may be misleading. Many risks may be excluded either by specific clauses or high excess, such as

- natural disasters and civil unrest;
- defective design, workmanship or materials;
- consequential losses such as financial losses arising from project delays or work stoppages; and
- testing and commission.

As demonstrated earlier, coverage is generally full insurance to market or replacement value and an escalator clause may be included.

Worker's compensation to workers below a certain salary for injury, disability or death is usually a statutory requirement, and the premium is about 1% of the payroll of these workers. The definition of "workers" excludes casual staff members of the contractor and employees of subcontractors.

Generally, a contractor is also liable for prosecution under the Workplace Safety and Heath Act and Regulations (or similar legislation) even if no injury occurs. Such penalties are generally uninsurable.

10.9 Bonds

A bond is treated here as a security or protection against certain actions in a contract, not as a financial instrument. Unlike two-party insurance (i.e. insurer and insured or policy holder), a bond involves three parties, namely, the client, contractor, and surety. The surety is typically a bank or an insurance firm, and sometimes the terms "bank bonds" and "surety (default) bonds" are used. In some contracts, there is a requirement for the issuer to be a local bank. If the contractor is an overseas contractor with no previous dealings with the local bank, he may need to approach an international bank he has dealings with that will then instruct the local bank to issue the bond against its counter-indemnity. This roundabout procedure is more expensive.

Unlike insurance, the premium for a bond is not treated as payment against risk but as a service fee to execute the bond. When a legitimate claim is made by the client, the surety pays up and then goes after the contractor for payment. This is why a contractor will always try to obtain a bond from a bank or insurance firm he is comfortable with, not a bank that will pay a client on demand (on production of certain documents to the bank) without his knowledge. An unfair call on bonds is a thorny issue, and conditional bonds provide better protection to the contractor than on demand bonds. In a conditional bond, certain conditions must be met before a claim for payment is entertained.

In the case of a joint venture between several contractors, each party would like to limit its bond liability to its share of the venture. Usually, this is agreeable to the bank or insurance firm. However, in cases where a partner is financially weak, all parties may be jointly and severally liable.

The various types of bonds commonly found in construction contracts are outlined below.

Tender or bid bond

A tender or bid bond of about 5–10 per cent of estimated contract price (the construction contract has not been awarded at tender stage) ensures that the contractor does not simply walk away from the project after being awarded the tender. A contractor may wish to walk away for reasons such as

- inability to cope with the workload because many of his bids for different projects are successful;
- discovery of a mistake in estimating the cost; or
- he is no longer interested in the work.

If a contractor walks away after submitting a winning bid, the penalty payable to the client is either the value as stipulated in the bond or the difference between the lowest two bids, whichever is lower. Instead of a bid bond, a tender deposit may also be used. The deposit is usually a bank check and the amount varies with contract value. Once the contractor has signed the contract, the check is returned to the contractor.

Performance bond

A performance bond of about 10 per cent of contract value and possibly higher (particularly in the US) is pledged as security for the general performance of the contractor. In the event of poor performance, default or insolvency, the surety has to pay damages to the client or hire another party to finish the work. The surety then has a right to reimbursement from the contractor. For his effort, the surety charges the contractor about one per cent of contract value as premium depending on contractor and project risks. The contractor includes this premium in pricing his bid.

An "on demand" performance bond is paid when called, as opposed on a "conditional" performance bond where proof of loss or damage (which may be disputed) is required. Even here, certain conditions must be met to prevent clients from calling bonds from a one-sided perspective. Generally, clients have to show that the contractor has failed to perform or has its contract terminated. For this reason, it is wise for the client to keep the surety informed of possible problems so that when they arise, calling the bond is less of a problem.

A performance bond is also a device to screen out risky contractors bidding for a project. A contractor who is unable to secure a performance bond may have financial troubles or poor track record.

Some project owners do not accept performance bonds. Instead, they prefer a bank letter of credit that pays cash when called on demand. Like performance bonds, there are also conditional letters of credit where the client needs to show that contractor has failed to perform. A letter of credit is generally cheaper than a performance bond. The premium is about one per cent of the contract amount covered by the letter of credit. For example, if the latter is 10 per cent of contract value, then the premium is $1\% \times 10\% \times$ contract value.

Payment bond

A payment bond or labor and material bond is sometimes used to guarantee that suppliers and subcontractors are paid by the main contractor and free the client's facility from liens lodged by subcontractors because of unpaid debts. It is usually not required if the contractor is financially sound and has a reputation for prompt payment.

Contractors may not pay subcontractors or suppliers because they are not paid by clients (e.g. because of poor demand for the project's output), resulting in severe liquidity problems for both parties. Unlike such "pay when paid" practices that tend to lead to unacceptable levels of subcontractor business failures, some countries have enacted Security of Payment Acts (e.g. New South Wales in 1999 and Singapore in 2004) to compel contractors to pay subcontractors and suppliers promptly (usually within four weeks) when work is done or goods and services are delivered without the need for expensive and lengthy arbitration or litigation. If he is unpaid, the subcontractor can make a claim to be paid within 10 days after which an independent adjudicator will decide on the outcome.

In the absence of such legislation, subcontractors and suppliers may claim the amount from the surety under the payment bond.

Another approach to contractor-subcontractor payment problems is to allow the market to decide, that is, for subcontractors to screen main contractors on their payment track record and financial standing. However, this method underplays the difficulties with screening contractors.

Advance payment bond

For large projects, the contractor is usually given an advance payment of up to 15 per cent of contract value to help with mobilization and other start-up costs.

An advance payment bond is required to assure the client that the contractor does not disappear after receiving a large sum of advance payment as start-up for the project.

Retention bond

Normally, a construction contract allows for retention of 10 per cent of each progress payment up to 5 per cent of contract value till the end of the defects liability period to ensure the contractor makes good all defects.

Retention can be problematic for contractors because the nature of "defects" can be subjective. If a project is losing money, clients may delay payment of retention to the contractor, pay a fraction (a 2.5% split on the retention sum is not uncommon), or not pay at all with a subsequent offer of a new contract for a different project.

In lieu of retention, contractors may prefer to issue a retention bond at the time the first retention on progress payment is made. Such a bond will automatically increase in value as further progress payments are made. Generally, clients are not in favor of such a bond because of possible problems in calling the bond (recall that the nature of "defects" is subjective) and, importantly, retention money is an important source of funds.

Bonding of subcontractors

The value of bonds is usually based on contract sum and therefore includes the value of subcontracting work. The main contractor will therefore need to obtain similar bonds from major subcontractors. If a subcontract fails to perform and this leads to a call on the main contractor's bond, the latter will make a claim on the subcontractor's bond.

Some subcontractors may not able to furnish a bond. It may be a genuine case of being a small business with little liquidity or that surety perceives the subcontractor as risky. Hence, the bond is a screening device as well.

Supply bond

A supply bond between a purchase and supplier guarantees purchaser that the supplier will supply the materials and items as contracted. If the supplier defaults, the surety will indemnify (compensate) the purchaser against the loss.

Questions

1 Use utility theory to explain why projects with higher stakes should be assessed based on higher hurdle rates.

2 What are the determinants of the risk premium?

3 Explain why partial insurance is common even though full insurance is optimal.

4 A contractor has two different "All Risks" insurance policies for two project sites that include damage to plant. A crane, while in transit from site A to site B, was damaged.

   a) How should the contractor claim for damages?
   b) How can the problem be resolved?

5 A simple example of a Contractor's All Risks (CAR) insurance policy for a contractor is given below. The premium for 10 employees is $1,000.

Item	Cover ($)
Contract works	200,000
Own plant and equipment	20,000
Hired plant	50,000
Drawings	10,000

Excesses:

Theft $1,000
Damage $1,000
Equipment $500

If you are a contractor and a broker has provided you with the above information, list the questions you would like your broker to answer before you make up your mind whether to purchase the policy.

6   In a conditional bond, certain conditions must be met before payment is made to the caller (client). Some of these conditions include

   •   consent by the contractor;
   •   certification by an independent third party; and
   •   an arbitrator's judgment in favour of the caller.

Discuss.

7   It is possible to purchase insurance on unfair calls on bonds. These are called bond risk insurance. The insurer will pay the amount of loss in an unfair call to the contractor, provided it is satisfied from the contractor's account that the call is unfair. The annual premium for such insurance is about 0.5 per cent of the value of the bond. Discuss the value of such insurance to the contractor.

8   Find the Taylor series approximation for $y = x^2$ at $x = 3$. Check if the approximation is good for $h = 0.1$.
    [$f(3 + h) = 9 + 6h + h^2 + \cdots$; $f(3 + 0.1) = 9.61$; the exact answer is $3.1^2 = 9.61$; approximation is very good]

11

Cash Flow Risks

11.1 Uncertain initial cost and cash flows

This chapter deals with risks associated with estimating the project initial cost and cash flows. Recall from Chapter 6 that if the net present value (NPV) criterion is used,

$$\text{NPV} = -C_0 + \frac{N_1}{1+r} + \frac{N_2}{(1+r)^2} + \cdots + \frac{N_n}{(1+r)^n} \tag{11.1}$$

where C_0 is initial cost, N_t is net operating income for year t, r is the discount rate, and n is the number of periods considered.

If project IRR(k) is used as the financial criterion to evaluate projects, then one solves for k in

$$0 = -C_0 + \frac{N_1}{1+k} + \frac{N_2}{(1+k)^2} + \cdots + \frac{N_n}{(1+k)^n}. \tag{11.2}$$

Lastly, the equity IRR (q) is computed from

$$0 = -E_0 + \frac{F_1}{1+q} + \frac{F_2}{(1+q)^2} + \cdots + \frac{F_n}{(1+q)^n} \tag{11.3}$$

where E_0 is initial equity, F_t is cash flow for year t, and n is the number of periods considered.

The discussion up to now has assumed that the initial cost and cash flows are certain. In practice, the initial cost and cash flows are affected by factors such as unexpected events, inadequate understanding of the business, insufficient data to make informed decisions, opportunism on the part of consultants, suppliers, and subcontractors, subjective judgments, statistical randomness, measurement errors, and linguistic imprecision.

The uncertainties in the initial cost and cash flows affect not only the project returns but also project liquidity and the ability to raise additional contingency funds. Hence, cash flow management is an important part of project management, and various methods of dealing with the uncertainties in cash flows are discussed below.

11.2 Estimating initial cost

Recall from Chapter 6 that the most accurate estimate of the initial cost of a project is the detailed estimate using priced Bills of Quantities (BQ) where quantities are accurately measured.

However, at the feasibility stage, the owner does not have elemental estimates based on the components (elements) of a facility (e.g. foundation, columns, walls, and so on) or detailed estimates based on detailed architectural drawings. In some procurement methods, the BQ is not used by the owner throughout the entire project, and the risk of an incorrect cost estimate is shifted to the contractor in submitting a bid. The owner locks in the project cost either through a guaranteed maximum price or the lowest bid. However, if the architectural drawings given to bidders are insufficiently detailed to allow for accurate estimation of bids, then contractors are likely to allow for large contingencies.

At the feasibility stage, the owner needs a preliminary cost estimate to determine if the project is viable. Without the architectural drawings to do elemental or detailed estimates, some other ways of estimating the project cost need to be found.

The most common method of approximate estimating is to use the *unit method* where the cost of a facility is the unit cost multiplied by the total area or units. For example, if the cost of a school is $100 psm and the total built-up area is 10,000 m^2, the total cost is $1,000,000.

The *factor technique* is an extension of the unit method. If the above cost estimate of a school excludes a swimming pool and running track, then these two features are added to $1,000,000 to determine the overall cost.

In the process industry, there are increasing returns to scale (see Section 7.3) so that

$$\frac{C_0}{C_E} = \left[\frac{Q_0}{Q_E}\right]^k$$

where C_0 is the initial cost of the proposed facility, C_E is the known cost of an exiting facility, Q_0 is the capacity of the proposed facility, and Q_E is the capacity of the existing plant. The coefficient k is usually in the range of 0.5 to 0.8. To account for inflation, an index of inflation (I_t) may be used so that the adjusted cost of the facility is

$$C_0 = C_E \left[\frac{Q_0}{Q_E}\right]^k I_t.$$

Hence, if the existing plant was built in 1995 and the inflation index was 105, then if the new plant is to be built in 2007 and the inflation index is estimated to be

120, then $I_t = 120/105 = 1.143$. If the new plant is 1.5 times the capacity of the existing facility, k is estimated to be 0.6 for this type of process facility, and the existing plant was built at a cost of \$100 m, then

$$C_0 = 100(1.5)^{0.6}(1.143) = \$146 \text{ m}.$$

Note that even after accounting for inflation, the cost of the proposed facility is still less than 1.5 times the cost of the existing facility.

11.3 Payback period

One simple way of dealing with the uncertainty in cash flows is to consider how long a project will pay itself back.

For example, if a project has an initial cost of \$10 m and an average net operating income of \$1 m per year, then

Payback period = $10/1 = 10$ years.

If the annual average net operating income of another project with the same initial cost is \$2 m per year, then

Payback period is $10/2 = 5$ years.

The project with a shorter payback period is preferred.

This method of dealing with uncertainty ignores the time value of money (i.e. no discounting is used) and tends to favor projects with shorter horizons, making it unsuitable for long-term infrastructure projects. In terms of Equation (11.2), if m is the payback period, the payback criterion sets

- $k = 0$ (i.e. no discounting);
- $N_{m+1}, \ldots, N_n = 0$ (i.e. periods beyond m are ignored);
- takes the average of $N_1, \ldots, N_m$; and
- solves for m.

For all its shortcomings, the payback period criterion has the advantage of simplicity as a quick method to screen projects before more detailed discounting methods are applied.

11.4 Conservative estimates

Another way of dealing with the uncertainties over project cash flows is to use more conservative estimates of net operating incomes rather than their expected values.

With a smaller numerator, the computed IRRs will be lower. This approach is often used in financing recreational facilities such as a stadium where demand is highly variable and seasonal. It is not advisable to be too optimistic about projected demand.

A related approach is the *certainty equivalent method* where each cash flow F_t is multiplied by λ_t where $0 < \lambda_t < 1$. More distant cash flows are assigned lower values of λ_t because they are less certain.

A downside with using conservative estimates of cash flows or the certainty equivalent method is that the firm may miss out on good opportunities because of its conservatism. Further, there is no clear guide on the level of conservatism or how to estimate λ_t. In the end, the techniques are somewhat judgmental or even arbitrary.

Recall from Chapter 2 that Gordon's formula is given by

$$V = C_1/(i - g)$$

where V is the present value of the asset, C_1 is the net annual income in the first year, i is the discount rate, and g is the constant growth rate in net annual income. It can be seen that V is sensitive to forecasts of i and g and, for this reason, it is relatively easy to over-estimate or under-estimate asset values. In particular, small differences in opinion on i and g can lead to very different views on project viability. A good example of this problem is the dot.com bubble of 1997–2001 where prices of technology stocks were grossly overvalued based on incorrect projections of earnings growth.

11.5 Risk-adjusted discount rate

If the net present value criterion is used, it is possible to use a higher risk-adjusted discount rate (r_a) to account for the perceived higher project risk. Then Equation (11.1) becomes

$$\text{NPV} = -C_0 + \frac{N_1}{1+r_a} + \frac{N_2}{(1+r_a)^2} + \cdots + \frac{N_n}{(1+r_a)^n}. \tag{11.4}$$

An example of a mark-up for risks is

$$r_a = r + \lambda_P + \lambda_C \tag{11.5}$$

where λ_P is the mark-up for project risk, and λ_C is the additional mark-up for country risk.

If IRRs are used, one marks up the hurdle rate rather than the discount rate to account for the risk (see Section 6.6).

11.6 Sensitivity analysis

There are various types of "what if" or sensitivity analyses that investigate what happens to the variable of interest (e.g. rate of return) if one variable such as the price of steel changes and other variables remain the same (i.e. held constant). If we make q the subject in Equation (11.3), then its functional form is

$$q = f(E_0, F_t, n). \tag{11.6}$$

Since E_0 and F_t are in turn functions of financial and project variables, Equation (11.6) may be written more generally as

$$q = g(x_1, \ldots, x_m). \tag{11.7}$$

Often, $g(.)$ is a nonlinear function, and the purpose of sensitivity analysis is to determine $\partial q/\partial x_i$, the rate of change in q with respect to each x_i, holding all other variables constant at mean values. Alternatively, one computes the change and impact in percentage (relative) terms. This is the elasticity of q with respect to x_i and it is given by

$$\text{Elasticity} = (\partial q/\partial x_i)(x_i/q).$$

A simple example will clarify the difference between the absolute and relative measures of impacts.

Example

Suppose $y = f(u, v) = 3u + 4v^2$. Then

$$\partial y/\partial u = 3; \text{ and}$$
$$\partial y/\partial v = 8v.$$

If the mean values of u and v are 5 and 10 respectively, then

$$\partial y/\partial u = 3; \text{ and}$$
$$\partial y/\partial v = 8v = 8(10) = 80.$$

We can write the above as

$$\Delta y = 3\Delta u$$
$$\Delta y = 80\Delta v$$

If $\Delta u = \Delta v = 1$, then a unit change in v clearly has a larger impact on y than a unit change in u.

In relative terms, the elasticities are

$$E_u = (\partial y/\partial u)(u/y) = 3(5/y) = 15/y; \text{ and}$$
$$E_v = (\partial y/\partial v)(v/y) = 8v(10/y) = 800/y.$$

Since $y = 3u + 4v^2 = 3(5) + 4(100) = 415,$

$$E_u = 15/415 = 0.04.$$
$$E_v = 800/415 = 1.93.$$

In words, a one per cent change in u leads to only a 0.04 per cent change (rise) in y. In contrast, a one per cent change in v leads to a 1.93 per cent change (rise) in y.

In sensitivity analysis, attention is paid to variables that cause major changes in q, that is, q is then said to be "sensitive" to changes in these variables. For example, one may vary the interest rate at 5, 10, and 15 per cent respectively. For each interest rate, a value for q is computed, giving a total of three values. A graph of q against interest rate may then be plotted to determine the slope, which is the required absolute measure of sensitivity.

The process is repeated by varying any other variable and holding the remaining variables constant. The sensitivity of q to each variable may then be plotted, and such a graph is called a spider plot. For example, Figure 11.1 shows how q varies with output price (x) and oil price (y) separately. If the output price rises by 5 per cent from its mean value, q is 13 per cent. Similarly, if oil price falls by 5 per cent from its mean value, q is 12 per cent. The slope is negative because if oil prices fall, q rises and vice versa. It can be seen that q is more sensitive to percentage changes in x than y because of the steeper slope.

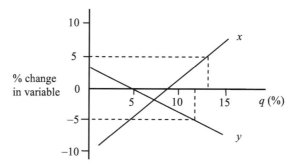

Figure 11.1 A spider plot.

A serious limitation of sensitivity analysis is that only one variable is varied at a time and its impact on q is ascertained. It does not consider the case where both variables x and y may change simultaneously. In such situations, their combined effect on q may be compounded or compensatory. In our example, a rise in oil prices will raise unit cost and this is likely to lead to a rise in output price. If the output is electricity, there is likely to be a reduction in electricity consumption following a price rise. We need to know the price elasticity of electricity demand to determine if the rise in electricity price and reduction in consumption will lead to a fall in total revenue.

11.7 Scenario analysis

Scenario analysis is another form of "what if" analysis. However, unlike sensitivity analysis, all variables are changed simultaneously in scenario analysis as different scenarios. The qualitative or quantitative impact of each scenario on q is then determined and corrective or preventive actions are then made.

If there are a large number of variables, there will be too many scenarios and the technique becomes cumbersome. Usually, only a small set of the more realistic scenarios is used and, for each variable, one estimates the optimistic value, most likely value, and the pessimistic value. For example, in the case of interest rates, 3 per cent may be the optimistic value, 5 per cent is the most likely value, and 7 per cent is the pessimistic value.

11.8 Monte Carlo analysis

Monte Carlo simulation extends scenario analysis by considering more than three values of a variable as well as the simultaneous change in all variables. In each trial, different values of each variable are selected based on assumed probability distributions and q is computed. By simulating a large number of trials, the probability distribution of q may be plotted.

Example

Consider a simple example where Monte Carlo simulation is used to determine the probability distribution of NPV where

$$NPV = -C_0 + \frac{N_1}{1+r} + \frac{N_2}{(1+r)^2} = -2000 + \frac{N_1}{1.10} + \frac{N_2}{1.10^2}.$$

It is assumed that the initial cost (C_0) and discount rate (r) are known with certainty and equal \$2,000 and 0.10 respectively. However, the net operating incomes N_1 and N_2 may vary simultaneously, and we wish to determine their combined impact on NPV.

The first step in the simulation is to determine the probability distributions of N_1 and N_2 and these are given in Table 11.1. For instance, N_1 takes values 900, 1,000, 1,100, and 1,200 (in thousand dollars) with probabilities 0.10, 0.20, 0.50 and 0.20 respectively. These probabilities are based on observed frequencies, experience or subjective judgment. The cumulative probability distribution is then computed and, as a check, the cumulative probability should sum to one. Finally, 2-digit (or 4-digit) numbers are assigned accordingly. In practice, 4-digit numbers are likely to be used and the first row will be assigned numbers 00–99, the second row will be assigned numbers 100–299, and so on.

Possible values	Probability	Cumulative probability	Assigned numbers
Variable N_1			
900	0.10	0.10	00–09
1,000	0.20	0.30	10–29
1,100	0.50	0.80	30–79
1,200	0.20	1.00	80–99
Variable N_2			
1,000	0.10	0.10	00–09
1,200	0.30	0.40	10–39
1,400	0.40	0.80	40–79
1,600	0.20	1.00	80–99

Table 11.1 Probability distributions and assigned numbers.

Similarly, N_2 takes on values 1,000, 1,200, 1,400, and 1,600 (in thousand dollars) with probabilities 0.10, 0.3, 0.4, and 0.2 respectively. The cumulative probability distribution is then computed and random numbers are assigned accordingly.

The next step in the simulation is to conduct repeat trials. In each trial, two random numbers between 00–99 will be drawn. Suppose the first two numbers are (23, 56). Then, using Table 11.1, 23 corresponds to $N_1 = 1,000$, and 56 corresponds to $N_2 = 1,400$. Hence, for the first trial,

$$NPV_1 = -2000 + \frac{1000}{1.10} + \frac{1400}{1.10^2} = 66.$$

In the second trial, suppose the numbers (9, 76) are drawn. From Table 11.1, 9 corresponds to $N_1 = 900$ and 76 corresponds to $N_2 = 1,400$. Then

$$NPV_2 = -2000 + \frac{900}{1.10} + \frac{1400}{1.10^2} = -25.$$

By running a large number of trials (e.g. 1,000), it is possible to plot the frequency or probability distribution of NPV. The mean and standard deviation of NPV may then be estimated for each project.

The random numbers such as (23, 56) for the first trial are generated using a computer command such as

INTEGER(100*RAND()).

The RAND() function generates a (reasonably) random number between 0 and 1. This is then multiplied by 100 to convert it to a number between 0 and 100, and the INTEGER function chops off the decimals. If 4-digit random numbers are desired, the computer command is changed to

INTEGER(10000*RAND()).

Monte Carlo simulation is conceptually easy to understand and not difficult to carry out. The key challenge is the ability to estimate the probability distributions of N_1 and N_2 with sufficient accuracy, particularly for distant cash flows or incomes. In such cases, one should allow sufficient variations in these variables to reflect the degree of ignorance.

In more sophisticated analyses, the net operating incomes or variables that affect them may be correlated. For instance, N_1 may be correlated with N_2. This is because a high value of N_1 in an economic boom is likely to lead to a high value of N_2 barring a sudden business downturn. Similarly, a low value of N_1 during a recession is likely to lead to a low value of N_2 in the next period. We can write

$$N_2 = \alpha + \beta N_1 + \varepsilon$$

where α is a constant (intercept), β is the slope and ε is the error term with zero mean and constant variance. Taking expectations,

$$E[N_2] = \alpha + \beta N_1 \tag{11.8}$$

since $E[\varepsilon] = 0$. The parameters α and β may be estimated using ordinary least squares (Figure 11.2). Then once N_1 is drawn in any trial, the value of N_2 is computed using Equation (11.8). This ensures that correlation is taken into account, that is, if a high value of N_1 is drawn, the corresponding value of N_2 is also high (see Figure 11.2). Similarly, if a low value of N_1 is drawn, the corresponding value of N_2 is also low.

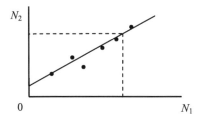

Figure 11.2 Correlated variables.

Some computer programs have made Monte Carlo simulation relatively easy to execute. For instance, instead of the probability distributions for N_1 and N_2 in Table 11.1, the software may allow one to specify a particular distribution such as a normal distribution with user specified mean and standard deviation. Similarly, if N_1 has a uniform distribution, then only the minimum and maximum values of N_1 need to be specified. For any probability distribution of a variable selected by the user, the software will automatically generate the cumulative probability distribution and perform the simulation.

How do we interpret the results of a Monte Carlo simulation? Figure 11.3 shows the mean and standard deviations of three different projects from a Monte Carlo simulation.

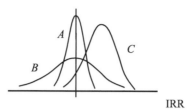

Figure 11.3 Choice of projects.

The choice between projects A and B is clear; since both projects have the same mean value, the project with the smaller standard deviation (risk) is selected or preferred (i.e. project A).

Although project C has a higher IRR than project A, it also has a higher risk. Hence, the choice between A or C depends on the hurdle rate as discussed in Section 6.6. Sometimes, the coefficient of variation (i.e. standard deviation/mean) is used and the project with the lowest coefficient is selected. This measure essentially scales the risk by the mean value.

11.9 Value at risk

After a series of spectacular financial disasters in the 1980s and 1990s, investors and lenders are paying more attention to the tail end of the distribution of their portfolio of assets or projects rather than just the expected return and standard deviation (Jorion, 2002). The discussion here will consider only value at risk (VaR) for a portfolio of projects.

If the area of the shaded region in Figure 11.4 is 5 per cent of the total area under the curve, there is a 5 per cent chance that, under normal market conditions, the value of the portfolio of projects may fall by more than 3 per cent a month (or any other suitable period). We say that the VaR for the $600 m portfolio of projects at 95 per cent confidence level is $(0.03)(600) = \$18$ m per month.

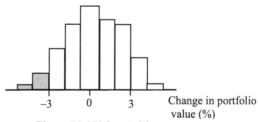

Figure 11.4 Value at risk.

VaR is based on historical data, and volatilities and correlations among asset returns may change. For a portfolio of projects, volatilities and correlations are difficult to ascertain because projects are unique one-off endeavors. Hence, the concept of VaR is more difficult to apply to projects.

11.10 Forecasting models

Clearly, one can reduce project risks by using better forecasts of project cash flows that, in turn, depend on revenues and costs.

In a naïve model, the quantity demanded for a project's output (D) is given by

$$D = \alpha + \beta t + \varepsilon \tag{11.9}$$

where α and β are population parameters to be estimated from a sample using ordinary least squares, t is time (e.g. quarterly or annually), and ε is the error term to account for

- unavoidable statistical randomness in D;
- errors in measuring D, such as using faulty meters; and
- variables omitted from the right hand side such as population growth or income.

The sample estimates of α and β in Equation (11.9) are the intercept "a" and slope of the line ("b") respectively (Figure 11.5). It is assumed that data are available from $t = 1$ to $t = T$, and the dotted lines beyond T are forecasts. The model is naïve because it uses only one variable (t) and it assumes the past will continue into the future in a fixed straight line (forecast A).

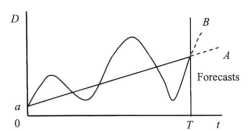

Figure 11.5 Naïve forecasting model.

Forecast B uses a flexible trend, called a moving average. It is more realistic than the straight line or fixed trend model. Hence, a dynamic autoregressive (AR) model such as

$$D_t = \alpha + \beta D_{t-1} + \gamma D_{t-2} + \varepsilon_t \tag{11.10}$$

tends to provide a better fit to the data. Here, D_t depends on its own previous values D_{t-1} and D_{t-2}. One could easily allow for more time lags but only two are shown here for simplicity. In practice, only a few lags are required.

The economic rationale for using lagged values of D is that the quantity demanded for the next period is closely related to previous levels of demand. For instance, next year's demand for electricity is closely related to this year's demand because consumers do not radically change their consumption of electricity. Further, the population of firms and households that purchases electricity does not change drastically within a year.

The same behavior can be said of stock prices; tomorrow's share price is closely related to today's closing price. The influence of D_{t-2} on D_t is likely to be much less than that of D_{t-1} because its effect is likely to have been incorporated into D_{t-1}. In essence, this is the *random walk hypothesis* for stock prices. The current share price incorporates all publicly available information about the profitability of the firm. There is little value in using historical prices or patterns to predict future stock prices.

If the error terms in Equation (11.10) are also lagged, we have an autoregressive moving average (ARMA) time series model

$$D_t = \beta D_{t-1} + \gamma D_{t-2} + \varepsilon_t - \lambda \varepsilon_{t-1}. \tag{11.11}$$

The lagged error terms are said to be "autocorrelated" because ε_t is correlated with ε_{t-1}. This effect may be seen in Figure 11.5. Observe that the portion of the curve above the trend line tends to persist, and the same can be said of the portion of the curve below the trend line. This means that a positive error in one period is likely to be followed by another positive error, and a negative error is likely to be followed by another negative one. Another way of thinking about the same effect is to realize that a boom is likely to persist for some time (e.g. five to ten years) before turning into a recession. Once the economy is in recession, it will persist for some time as well. If there is no autocorrelation, the curve will alternate frequently (randomly) above and below the trend line.

A limitation of Equation (11.11) is that a time series cannot be satisfactorily modeled in this way unless the series is (weakly) stationary. This implies that the mean value and variance of the series do not change with time. If a series is trending up, as in Figure 11.5, the mean is no longer constant. To see this, if we divide the series into two periods (1 to $T/2$, and $T/2 + 1$ to T) and compute the means separately, they will differ substantially. The second period of the series in Figure 11.5 is also more volatile, indicating that the two periods do not have the same variance as well.

However, all is not lost. A simple solution is to transform the original series D into a new stationary series q using

$$q_t = D_t - D_{t-1}. \tag{11.12}$$

This technique is called first-order differencing or simply differencing. A series with a linear trend such as the one shown in Figure 11.5 will be stationary after differencing. If the trend is exponential (this is rare in practice for economic series), it is necessary to take logarithms of the original series to transform it into a linear trend before differencing. Note we have used the symbol D to refer to a variable as well as a time series. The latter is sometimes written more fully as $\{D_t\}$ or $\{D_t,\ t = 1,...,\ T\}$. It should be clear from the context whether D refers to a variable or time series.

If the original series has a linear time trend and q is the differenced series, the transformed model is

$$q_t = \alpha_1 q_{t-1} + \alpha_2 q_{t-2} + \varepsilon_t - \lambda\varepsilon_{t-1}. \tag{11.13}$$

This is called an autoregressive integrated moving average (ARIMA) model (Box and Jenkins, 1970). It may be written as ARIMA (m, d, s) where m is the number of lags in q, d is the order of integration or number of times a series is differenced to make it stationary, and s is the number of lags in ε. Equation (11.13) is an ARIMA (2, 1, 1) model. Note the parameters of this model cannot be estimated using ordinary least squares because of the presence of the autocorrelated error terms. Nonlinear least squares estimation is often used in such cases, and this is beyond the scope of this book.

In practice, ARIMA models are rarely used to forecast project demand because they are also naïve, that is, they do not take into account the impact of other variables. A logical extension of the model is to include more independent variables, such as

$$q_t = \alpha_0 + \alpha_1 x_t + \alpha_2 p_t + \alpha_3 g_t + \varepsilon_t - \lambda\varepsilon_{t-1}. \tag{11.14}$$

Here, p (output price) and g (population growth) are additional exogenous variables that may potentially affect q. As before, only two variables are shown to keep the exposition simple. An exogenous variable such as x is determined outside the system, and an endogenous variable (q in Equation (11.14)) is determined by the system. A simple example is the function $y = f(x)$. Here x is determined outside the system (equation), and once x is selected, the endogenous variable y is then computed using $y = f(x)$.

What are the weaknesses of the multiple regression model in Equation (11.4)? Apart from the problem of autocorrelation discussed above, we need to forecast the values of the exogenous variables p and g first, substitute them into the right hand side of Equation (11.4), and then forecast the endogenous variable q. If there are many such exogenous variables, the accuracy of the forecasts is unlikely to be high because one is using forecasts of these variables just to forecast the value of the endogenous variable q.

Further, is the variable p in Equation (11.14) really exogenous? To better understand the issue, consider a simple model

$$q_t = \alpha + \beta p_t + \varepsilon_t. \tag{11.15}$$

As before, q is quantity demanded, p is output price, and ε is the error term. If p is exogenous, it is determined outside the equation. If we ignore the error term ε, then $q = \alpha + \beta p$ is the equation of the demand curve in Figure 11.6.

If ε is now added, it represents shifts in the demand curve (as shown by the doubled-headed arrow). If ε is positive, the demand curve shifts outwards; if it is negative, the curve shifts inwards.

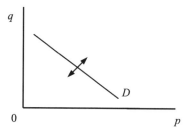

Figure 11.6 Shifts in demand curve.

Consider what happens when a supply curve S is added (Figure 11.7). If ε is positive, D shifts outwards and the price p has increased as well. This means that a positive value of ε raises the value of p. Conversely, a negative value of ε lowers the value of p. Thus, rather than being independent, what we have argued is that ε and p are correlated.

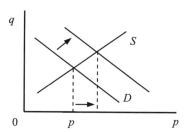

Figure 11.7 Correlation between p and ε.

What are the implications of this correlation? Intuitively, it means that q is not determined solely by the demand equation (i.e. Equation (11.14) or the simpler Equation (11.15)) because p is also an endogenous variable. If Equation (11.5) is estimated using ordinary least squares, the estimate of β will not be consistent, that is, our estimator b will not be centered on β even if a large sample is used. This bias is undesirable.

The proof of the inconsistency of b as an estimator of β is not difficult if the matrix approach is used. Recall from Equation (5.22) that the normal equations may be written as

$$
\begin{aligned}
\mathbf{b} &= (\mathbf{X}^T\mathbf{X})^{-1}\mathbf{X}^T\mathbf{y} \\
&= (\mathbf{X}^T\mathbf{X})^{-1}\mathbf{X}^T(\mathbf{X}\beta + \varepsilon) \\
&= \beta + (\mathbf{X}^T\mathbf{X})^{-1}\mathbf{X}^T \, \varepsilon.
\end{aligned}
$$

In ordinary least squares regression, $\mathbf{X}$ is assumed to be fixed (exogenous) and uncorrelated with the error term ε. Then, taking expectations on both sides give

$$
E[\mathbf{b}] = \beta + E[(\mathbf{X}^T\mathbf{X})^{-1}\mathbf{X}^T \, \varepsilon] = \beta. \tag{11.16}
$$

Now if $\mathbf{X}$ and ε are correlated,

$$
E[(\mathbf{X}^T\mathbf{X})^{-1}\mathbf{X}^T \, \varepsilon] \neq 0
$$

even for large samples. This means that $E[\mathbf{b}] \neq \beta$ and the estimator $\mathbf{b}$ is inconsistent.

A solution to this problem of inconsistency is to recognize that if p is determined by demand and supply, a supply equation needs to be added to Equation (11.14) and the entire system of simultaneous equations is then solved using special techniques. These techniques are required because it may not be possible to obtain unique estimates of the parameters (the identification problem) or there are too many parameters in the system of equations.

Traditionally, a priori restrictions such as $\alpha_3 = 0$ or $\lambda = 0$ in Equation (11.14) were used to reduce the number of parameters to a manageable number. Sims (1980) has criticized these a priori restrictions as "incredible" (i.e. unreal) and proposed treating all variables as endogenous to be determined within the system of structural equations. This is the vector autoregression (VAR) model (not to be confused with VaR in Section 11.9), and an example is

$$
\begin{bmatrix} y_{1t} \\ y_{2t} \end{bmatrix} = \begin{bmatrix} \alpha_1 \\ \alpha_2 \end{bmatrix} + \begin{bmatrix} \pi_{11} & \pi_{12} \\ \pi_{21} & \pi_{22} \end{bmatrix} \begin{bmatrix} y_{1t-1} \\ y_{2t-1} \end{bmatrix} + \begin{bmatrix} v_1 \\ v_2 \end{bmatrix}. \tag{11.17}
$$

In this simple VAR model, there are only two variables, y_1 and y_2, and both are endogenous. Even with a small set of variables, there are many parameters to estimate (i.e. two αs, four πs, and variances and covariances of the error terms). As before, the VAR process is assumed to be weakly stationary. If this is not the case, each series must be differenced first. In vector form, we can write Equation (11.17) as

$$
\mathbf{y}_t = \alpha + \pi\mathbf{y}_{t-1} + \mathbf{v}_t. \tag{11.18}
$$

More generally, if the vector $\mathbf{y}_t$ has more than two variables and additional lagged terms are included, then

$$y_t = \alpha + \pi_1 y_{t-1} + \cdots + \pi_p y_{t-p} + v_t. \qquad (11.19)$$

Unfortunately, even with a large VAR model that treats all variables as endogenous, there is the serious objection that the parameters may not be stable if Equation (11.19) is used to evaluate policies. This is because economic agents are not passive and have *rational expectations* that anticipate the impacts of a particular policy (Lucas, 1976). This means that they will react accordingly and change their behavior. If the parameters are unstable, predictions from VAR policy models will not be accurate.

This brief overview of forecasting models from naïve trend fitting to large scale simultaneous equations models or VAR models ignores many of the technical and complex details in estimating such models. It is intended to provide an intuitive grasp of the strengths and weaknesses of different types of forecasting models. It seems clear that large models are not necessarily superior to simpler models, and naïve models are unlikely to predict well.

Questions

1   The use of a risk-adjusted discount or hurdle rate to evaluate projects is sometimes objected on several grounds.

a) The approach assumes capital markets are perfect and capital is mobile so that risk-adjusted rates of returns across investment classes are equal.

b) It confuses the lender's risk and project risk.

c) Generally, infrastructure projects are initially risky but once the facility is built, the risks are relatively low. Hence, the use of a constant risk mark-up or premium across all discounting periods is not sound.

d) The risk mark-up is subjective.

e) Usually an aggregate measure to account for risk is unwise. The impact of each variable on the project should be determined separately.

f) There is no adjustment for capacity utilization.

g) It neglects growth options. Risky projects may have tremendous upside growth potential. One should defer the project until such time rather than use a risk-adjusted discount rate.

h) Management can often reduce project risk through diversification and other measures. Viewed in isolation, a project may appear risky. However, viewed as a portfolio of projects, the risk may be much lower.

i) Projects may be executed in phases. If Phase I is unsuccessful, subsequent phases will be aborted, and considerable uncertainty disappears before each phase is started.

j) Projects may be undertaken for strategic and other reasons.

Evaluate these claims.

2   Show that the risk-adjusted discount rate (r_a) can be written as

$$r_a = r_F + \pi + \lambda$$

where r_F is the risk-free rate, π is the expected rate of inflation, and λ is the risk premium. [Hint: Use $(1 + r_a) = (1 + r_F)(1 + \pi)(1 + \lambda)$]

3   Many firms forecast a "worst case" scenario to ensure that the company has sufficient funds to repay debt.

a) What are the limitations of using a "worst case" scenario?

b) If the firm has cash flow problems, what can it do?

4   What are the strengths and weaknesses in using Monte Carlo simulation to analyze project risk?

5   Some project managers put a lot of faith in forecasts generated by large-scale models. Give reasons why such faith may be misplaced.

6   Consider the short-run dynamic model

$$Y_t = \alpha + \beta X_t + \lambda X_{t-1} + \varepsilon_t.$$

If X and Y are not stationary, one could consider a differenced model

$$y_t = \gamma x_t + \phi x_{t-1} + v_t.$$

In using a differenced model, we lose information on the relation between X and Y in the first equation because, in the long run, $X_t = X_{t-1} = X^*$ and

$$Y^* = \alpha + \beta X^* + \lambda X^* = \alpha + (\beta + \lambda)X^*$$

where Y^* and X^* are long-run equilibrium values. It is possible that, from the first equation,

$$\varepsilon_t = Y_t - \alpha - \beta X_t - \lambda X_{t-1}$$

is stationary even though X and Y are not stationary (i.e. trending). We say that X and Y are cointegrated. How useful is such a concept in a predictive or explanatory model?

12

Financial Risks

12.1 Derivatives

This chapter deals with instruments for managing financial risks. The main instruments to be discussed here are derivatives. A derivative is a contract between two parties that derives its value from some underlying asset price, index, or reference interest rate. Derivatives include forwards, futures, swaps, caps and floors, and options.

12.2 Forwards

In a forward contract between two parties, an item (e.g. currency, commodity or product such as oil) is delivered at a specified quantity, standardized quality, price, and future date. At the delivery date, a trade must occur.

For the buyer, a forward contract locks in the price of the item now rather than the prospect of higher prices in future. For the seller, it is also a means to set the price he will get now rather than face the prospect of lower future prices. Both parties are said to hedge price risk.

A forward contract is not marketable, that is, it cannot be sold to a third party privately or in an organized exchange. Both parties are therefore committed to the deal and it can be cancelled only by mutual agreement. This makes forward contracts less attractive than futures contracts that can be traded in an organized exchange.

12.3 Futures

As noted above, a futures contract is similar to a forward contract except that it can be traded in an organized exchange. In turn, this requires that futures contracts (or simply futures) be standardized. The buyer is said to be on a long hedge, and the seller is on a short hedge.

Futures may also be based on a stock index. If I hold a portfolio of 10 stocks currently valued at $100,000 and feel that the stock market index (currently at 2,000 points) is going to fall in the short term, it may not be worthwhile to sell my entire portfolio because of the high transaction costs. Instead, I may trade in a stock index futures contract after finding a broker and putting an initial (margin) deposit. The value of a futures contract is given by

$$V = \text{unit value} \times \text{index value} = 100 \times \text{index value}$$

where unit value is fixed by the exchange (shown as $100 above), and index value is the prevailing value of the stock market index (currently 2,000 points). Hence, if I sell one futures contract, its worth is

$$V = 100 \times 2,000 = \$200,000.$$

Buying and selling futures contracts are similar to buying and selling apples. If the market index falls to 1,900 after three months, I made a loss on my stock portfolio. However, I will make a gain if I buy a futures contract at 1,900 points. Excluding brokerage, my gain on the futures contracts is as follows:

> Now: Sell one futures contract at 2,000 points for $200,000
> Three months later: Buy one future contracts at 1,900 points for $190,000
> Gain: $10,000

On the other hand, if the market index rises to 2,200 points instead of falling to 1,900 points, my loss on the futures contract (excluding brokerage) is as follows:

> Now: Sell one futures contract at 2,000 points for $200,000
> Three months later: Buy one future contracts at 2,200 points for $220,000
> Loss: $20,000

Hence, it does not matter whether the stock market index moves up or down; overall, my position is hedged where a win in one market is partially offset by a loss in the other market. In the above example, I made money by selling a futures contract high and buying it back low. The other way to make money is to buy it low and sell it high.

As noted earlier, trading in indexed futures contracts is similar to that of buying and selling of shares except that, in the former, one is not buying or selling equity but contracts. Hence, no dividend is payable on these contracts. However, as we have seen, a futures contract is used primarily for hedging risks.

A futures market serves another important function, and that is to help discover prices. It is possible to ascertain or "discover" the expected market price of an item (such as a barrel of oil) in a futures market as reported in the newspapers or exchanges.

12.4 Swaps

An interest rate swap provides a mechanism for a borrower to reallocate exposure to interest rate fluctuations by swapping interest payment obligations with another party.

In Figure 12.1, party *A* agrees to pay party *B* periodic interest payments at LIBOR + 1 per cent (recall that LIBOR is the London Interbank Offered Rate) in exchange for a fixed 4 per cent swap rate at an agreed currency. If LIBOR at the end of next month (the agreed date) is 2 per cent (say), *A* then pays 3 per cent interest to *B*, and *B* pays the fixed 4 per cent interest to *A*. In practice, only net payments are made, that is, the 1 per cent difference in interest payment on a previously agreed (notional) principal payable from *B* to *A*. On the other hand, if LIBOR is 5 per cent, then *A* pays *B* the difference of 2 per cent interest on the principal.

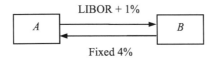

LIBOR + 1%

Fixed 4%

Figure 12.1 Interest rate swap.

For a swap to be mutually beneficial, both parties must have different expectations on future LIBOR rates. In other instances, *B* may be a mortgage lender who receives periodic interest payments at fixed rates and offers floating rates to its depositors. By lending on long term to borrowers and borrowing on short term from depositors, *B* is substantially exposed to interest rate fluctuations. It therefore seeks a party willing to swap its fixed rate exposure with floating rates.

The swap in Figure 12.1 may appear less confusing if one examines it entirely from the perspective of one party. For instance, party *A* is swapping its floating rate obligations for a fixed rate obligation. This allows *A* to "lock in" its interest rate obligations at the fixed rate of 4 per cent.

A swap contains counterparty risk, that is, one party may not honor its obligation. Usually, a bank acts as an intermediary to arrange the swap transaction for a fee. It will also assume the credit risk of both parties in the event of a default by either party.

A currency swap works in a similar manner with a bank acting as an intermediary. Both parties will exchange, from the beginning, a principal sum denominated in one currency for another at a specified exchange rate. They will reverse the transaction at the same exchange rate on a predetermine date. Meanwhile, they will cover each other's interest commitments during the term of the swap.

In Figure 12.2, party *U* is a US firm with comparative advantage in raising a US$10 m loan, that is, it can borrow at a lower rate of interest. Party *S* is a Singaporean firm with comparative advantage in raising a S$16 m loan. Both parties agree to swap their loans at the spot exchange rate of 1US$ to S$1.60. *S* will pay interest to the lender on the US$10 m loan, and *U* will similarly pay interest on the S$16 m loan. These interest payments are shown as dashed arrows in Figure 12.2. At the end of the swap period (e.g. two years later), both parties will reverse the transaction.

A swap is a private transaction and is an off-balance sheet item. Since the minimum swap period is often two years, it is used to manage medium to long term capital needs. Banks now provide fairly standard swap agreements (e.g. the 2002 International Swaps and Derivatives Association Master Agreement) with early termination clauses.

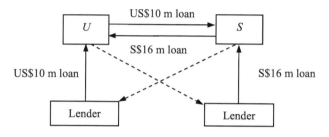

Figure 12.2 Currency swap.

12.5 Caps and floors

An interest rate cap protects the buyer from interest rate fluctuations above a certain cap rate. For instance, if I buy a 5-year interest rate cap at a certain premium, I will be paid at the end of each month if the reference interest rate (e.g. LIBOR) rises above the cap rate. If LIBOR falls below the cap rate, nothing is paid. The payment is based on the difference between the reference rate and the cap rate on a notional principal.

The premium depends on market conditions. If interest rates are expected to rise during the term of the loan, the bank is likely to raise the premium for an interest rate cap.

An interest rate floor works the other way. The buyer is paid an amount equal to the difference between an agreed floor rate and LIBOR should the latter fall below the floor rate. It protects the buyer against interest rate declines. As before, the principal is notional.

An interest rate "collar" is a combination of an interest rate floor and cap. It protects the buyer against interest rate rises and declines outside the collar (i.e. cap and floor rates).

Example

John borrows $10 m for a term of 10 years from a bank at LIBOR + 1 per cent. If LIBOR is currently 5 per cent, John is effectively paying 6 per cent interest, and does not wish for LIBOR to go beyond 7 per cent. To do so, John needs to pay the bank a premium for undertaking the interest rate risk.

However, John can devise an interest rate collar to reduce the premium. He can offer the bank a floor of 4 per cent for LIBOR, that is, John will compensate the bank if LIBOR falls below 4 per cent (see Figure 12.3). An interest rate collar therefore provides a mechanism for both parties to share interest rate risk.

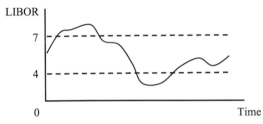

Figure 12.3 An interest rate collar.

12.6 Real options

Projects may contain options such as the option to buy a piece of land, to expand production into an adjacent site, or to take further part if an oil exploratory project or research and development project is successful. These real options differ from financial options because they involve real assets.

A developer who buys a piece of land may have the option to buy an adjoining site owned by the same seller. For the developer, buying both tracts of land at the onset is costly and risky. If the first project does not do well, he is stuck with many unsold units. Hence, it is advisable to buy only the first site at market value and have a call option (or simply "call") to buy the second site at a pre-determined price (called the strike price) at a future date (date of expiry).

If the first site is bought at $100 per m², the developer may buy a call on the second site at $105 per m² to be exercised in two years' time (Figure 12.4). If the first project does not sell well, the developer will not exercise the option at the date of expiry ($t + 2$). Since the land owner cannot sell the second site within two years because of the option, he will demand to be paid an option price. This is why the pricing of options is important. Further, options are more attractive than futures because only a relatively small amount of money (the option price) is involved to obtain an option.

A second reason why option pricing is important is that if the NPV or IRR criterion is used to evaluate the financial viability of projects, it is possible that a project may not be viable at this moment because of current market conditions. However, with real options, a project that is initially not feasible may become viable when more information is known about future cash flows. Hence, there is value in waiting for more information before one decides to commit further resources into a project.

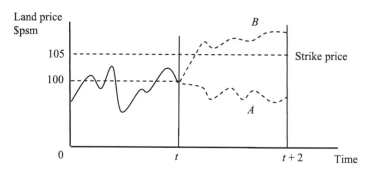

Figure 12.4 The pricing of options.

Example

Explain how a poker game may be viewed as an option.

In a poker game, one is given the choice to follow, raise the stake, or exit from the game. Following or raising the stake entails committing some resources (option price) to obtain the right to receive the next card (i.e. additional information). Clearly, if one has a good hand and is confident of winning, then raising the stake is a good strategy.

There are many types of real options in development projects. Some examples include

- a straight option (or plain vanilla option with no frills) to buy a piece of land;
- a rolling option contingent on development success in the initial phase;
- a price escalation option where the purchase price rises with the exercise date to encourage early exercise;
- a lease and release option where the land is first leased to the developer before it is released at an agreed price; and
- a rental purchase agreement where part of the rent forms the down payment when the land is released.

Suppose the price of the call option in Figure 12.4 is $10 per m². At the expiry date, the total land cost is $10 + $105 = $115 per m². If the market price of land falls below this value (trajectory *A*), the developer is better off buying land from the open market and not exercise the option.

However, if the market price of land rises above $115 per m² (trajectory *B*), the developer will exercise the option. Intuitively, the price of the call option (*C*) depends on

- the current land price, S ($100 per m^2);
- the exercise or strike price, K ($105 per m^2);
- the time to expiry, T (2 years);
- the volatility of returns on land prices, σ; and
- r, the risk-free interest rate.

Note r is the risk-free interest rate; we have departed from using r_F as the risk-free rate for notational convenience.

The current land price provides a benchmark for predicting the strike price and, the longer the time to expiry, the more difficult it is to predict the strike price. In turn, the prediction is affected by the volatility of price fluctuations. If land prices are volatile, options are more valuable if one could exercise it at any time before expiry. There is value in waiting for land prices to fall before exercising the option. Finally, the risk-free interest rate affects the price of a call option because future values need to be discounted to present value.

So how is the price of the call determined? Since option pricing models were first developed in the stock market, we shall begin with share prices and then apply the model to land prices.

12.7 The binomial model

Consider a simple binomial tree (Figure 12.5) where the current stock price ($75 per share) either moves up to $100 or down to $50 in the next period (or date of expiry), say one year later. Assuming no taxes, dividends, transaction cost, or trade restrictions during the life of an option, what is the price of a call?

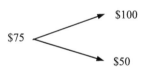

$75

$100

$50

Figure 12.5 A binomial tree.

As a first approximation, the price of the call can be determined by finding the expected value of the share price at the date of expiry and discount it to present value using the risk-free interest rate. If the share price has an 80 per cent chance of going up and a 20 per cent chance of going down, its expected value is

$$E = 0.8(100) + 0.2(50) = \$90.$$

The expected price of the call at time of expiry is

$$C = \$90 - \$75 = \$15.$$

If the risk-free interest rate is 3 per cent, the present value of the call is $15/1.03 =$ $14.56.

Suppose I buy a share of the stock at $75 and sell two calls to you for $14.56. My initial outlay is

$$I = 75 - 2(14.56) = \$45.88.$$

In the next period, the value of my portfolio (V_U) if the share price moves up to $100 is

$$V_U = \text{stock value} + \text{call loss}$$
$$= 100 + 2(75 - 100) = \$50.$$

If the share price moves down to $50, the value of my portfolio is

$$V_D = \text{stock value} + \text{call loss}$$
$$= 50 + 0 = \$50.$$

The call loss is $0 because you will not exercise the option if the price falls to $50. Note the value of my portfolio is $50 irrespective of whether the share price moves up or down.

On the other hand, my initial outlay is only $45.88. I have made $50 - 45.88 =$ $4.12 regardless of what happens to the share price in the next period. Clearly, if I can devise a sure-win portfolio, so can any other discerning investor. This means that it is not an equilibrium position.

More generally, suppose I buy x shares at $75 per share now and sell y calls at C per option with a strike price of $80 per share. Then

$$\text{Value of portfolio} = 100x - 20y \quad \text{if share price is \$100}$$
$$= 50x \quad\quad\quad\quad \text{if share price is \$50}$$

If the share price is $100, the value of the x shares is $100x$, and because the strike price is $80 (i.e. below $100), the y options will be exercised giving a loss of ($100 - $80)$y = \$20y$.

If the share price falls to $50, the value of the x shares will be $50x$. Since $50 < \$80$, the y options will not be exercised and end up worthless. Hence the value of my portfolio is only $50x$.

The portfolio is said to the risk-free if its value is independent of whether the share price moves up or down. Then

$$100x - 20y = 50x$$

or

$$y = 5x/2.$$

The original cost of my portfolio is $75x - (5x/2)C$. Note the sign is negative because I sold options, thereby reducing my cost. The value of the portfolio in the next period is $50x$ (= $100x + 20y$) irrespective of whether the share price moves up or down. Hence,

$$\text{Portfolio gain} = 50x - [75x - (5x/2)C].$$

Of course, if I can make money out of it, someone else could as well. Hence, if there are no arbitrage opportunities, the portfolio gain is zero so that

$$0 = 50x - [75x - (5x/2)C]$$

or

$$C = \$10.$$

This is the price of the call option, and we can discount it to present value as before.

The above example is useful in explaining how a call is priced using arbitrage. However, it is unrealistic in the sense that it considers only two possible outcomes (\$100 or \$50). In practice, share prices can take any value above zero. Hence, the next step in our understanding of real options is to generalize the simple binomial model to the continuous case where future share prices can take any non-negative value rather than just two outcomes. The theory of option pricing then becomes more complicated but the basic ideas remain the same.

12.8 Stochastic processes

Obviously, the first step in option pricing is to build a suitable model of stock price movement. If

$$y = f(x) = 2x,$$

then $f(.)$ is called a deterministic function because given x, it is possible to compute y exactly. Unfortunately, stock prices are not deterministic; rather, they are stochastic. To understand this, consider the simple stochastic model

$$x_t = x_{t-1} + u_t$$

where x is stock price, t is time, $t = 1,..., T$, and u is a random error term. Note that the second equation is time ordered and contains a random term. This means that x_t cannot be determined exactly even if x_{t-1} is known. An example will clarify.

Example

Simulate the stochastic process above using a fair coin.

For simplicity, suppose $x_1 = 10$ and u_t takes the value of $+1$ if the toss is a head and -1 if it is a tail. Then if the first toss is a head,

$$x_2 = x_1 + u_2 = 10 + 1 = 11.$$

If the second toss is also a head, then

$$x_3 = x_2 + u_3 = 11 + 1 = 12.$$

If the third toss is a tail, then

$$x_4 = x_3 + u_4 = 12 - 1 = 11,$$

and so on.

If we plot the movement of $\{x_t\}$, it looks something like that of Figure 12.6. Obviously, even for the same experiment, each person will obtain a different trace depending on how the coin lands.

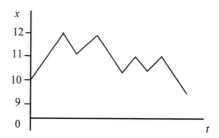

Figure 12.6 A simulated stochastic process.

12.9 The Wiener process

If the time period in Figure 12.6 is reduced to smaller periods, the trace will become smoother, that is, the process becomes more continuous. A popular model of this process is the Wiener process or Brownian movement. It is given by

$$dz = \varepsilon\sqrt{dt} \qquad\qquad (12.1)$$

where dz is the change in stock price over a small time interval dt and $\varepsilon \sim N(0, \sigma^2)$. Thus,

$$E[\varepsilon] = 0; \text{ and} \qquad\qquad (12.2a)$$
$$\text{Var}(\varepsilon) = \sigma^2 . \qquad\qquad (12.2b)$$

From Equation (12.1),

$$E[dz] = E[\varepsilon\sqrt{dt}] = \sqrt{dt}\, E[\varepsilon] = 0; \qquad\qquad (12.3)$$
$$\text{Var}(dz) = \text{var}(\varepsilon\sqrt{dt}) = dt\, \text{var}(\varepsilon) = \sigma^2 dt. \qquad\qquad (12.4)$$

These two equations account for the popularity of the Wiener process. Since $var(dz) = dt$, its volatility increases proportionately with time. To see why this process is also called a Brownian movement, let z_t be the position at time t and we take a small step Δz either left or right with equal probability. Then

$$z_t = \Delta z(z_1 + z_2 + \cdots + z_k)$$

where k is the number of steps taken and

$z_i = +1$ if the ith step is to the right; and
$\quad = -1$ if it is to the left.

If the probability of stepping left or right is the same $(p_i = p = 0.5)$, then

$$E[z_i] = \Sigma\, p_i z_i = 0.5(1) + (0.5)(-1) = 0,$$

and

$$
\begin{aligned}
Var(z_i) &= E[z_i - E(z_i)]^2 &&\text{(by definition of variance)}\\
&= E[z_i - 0]^2 &&\text{(since } E[z_i] = 0)\\
&= \Sigma\, p_i z_i^2 = 0.5\Sigma\, z_i^2 = 0.5[1^2 + (-1)^2] = 1.
\end{aligned}
$$

Hence,

$$
\begin{aligned}
E[z_t] &= 0;\\
Var(z_t) &= var[\Delta z(z_1 + z_2 + \cdots + z_k)]\\
&= (\Delta z)^2 var(z_1 + z_2 + \cdots + z_k) &&\text{(since } var(cx) = c^2 var(x) \text{ if } c \text{ is a constant)}\\
&= (\Delta z)^2 [var(z_1) + \cdots + var(z_k)] &&\text{if the } zs \text{ are independent}\\
&= k(\Delta z)^2.
\end{aligned}
$$

In the time interval t, $k = [t/\Delta t]$ where [.] denotes the largest integer. Hence,

$$Var(z_t) = t(\Delta z)^2/\Delta t.$$

If we now let Δz and Δt tend towards zero, the variance will tend towards zero. To prevent this collapse of the variance, let

$$\Delta z = \sigma\sqrt{\Delta t}$$

for some positive constant σ. Then

$$Var(z_t) = t(\sigma\sqrt{\Delta t})^2/\Delta t \rightarrow \sigma^2 t$$

as $\Delta t \rightarrow 0$. Since $z_t \sim N(0, \sigma^2 t)$,

$$
\begin{aligned}
E[dz] &= 0, \text{ and}\\
Var(dz) &= \sigma^2 dt.
\end{aligned}
$$

These are precisely Equations (12.3) and (12.4). If σ is set to one, then

$$\text{var}(dz) = dt$$

and rescaling may be made simply by multiplying dz by σ.

12.10 The generalized Wiener process

The movement of a stock price S may be modeled as a generalized Wiener process

$$dS/S = \mu dt + \sigma dz \qquad (12.5)$$

or

$$dS = \mu S dt + \sigma S dz. \qquad (12.6)$$

Here μ is called the drift parameter, and dz has been scaled by σ. This means that $\text{var}(dz) = dt$.

Ignoring the last term in Equation (12.5) for the moment and integrating both sides, we get

$$\int dS/S = \int \mu dt,$$

i.e.,

$$\log(S) = \mu t + c \qquad (c = \text{constant of integration})$$

or

$$S = S_0 e^{\mu t} \qquad (12.7)$$

where S_0 is the initial stock price. Thus, from Equation (12.5), stock price is modeled to drift exponentially as in Equation (12.7) plus a scaled stochastic term σdz to account for the wavy features in Figure 12.7.

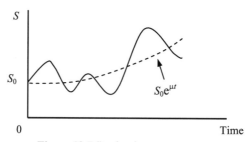

Figure 12.7 Stock price movement.

To proceed further, assume that there are no transaction costs, taxes, trade restrictions, or dividends during the life of the option and unlimited borrowing at risk-free rate r is available. Ito's lemma is also required to link the stock and option prices, and this is discussed in the next section.

12.11 Ito's lemma

If stock price S is defined by Equation (12.6) and the option price $F = F(S, t)$ depends on two variables S and t, then

$$dF = \left[\mu \frac{\partial F}{\partial S} + \frac{\partial F}{\partial t} + \frac{1}{2} \sigma^2 S^2 \frac{\partial^2 F}{\partial S^2} \right] dt + \sigma S \frac{\partial F}{\partial S} dz. \qquad (12.8)$$

Proof

The Taylor series approximation for $F(S, t)$ is

$$dF = \frac{\partial F}{\partial S} dS + \frac{\partial F}{\partial t} dt + \frac{1}{2} \left[\frac{\partial^2 F}{\partial S^2} (dS)^2 + 2 \frac{\partial^2 F}{\partial S \partial t} (dSdt) + \frac{\partial^2 F}{\partial t^2} (dt)^2 \right]$$

$$\approx \frac{\partial F}{\partial S} dS + \frac{\partial F}{\partial t} dt + \frac{1}{2} \frac{\partial^2 F}{\partial S^2} (dS)^2, \qquad (12.9)$$

ignoring higher order terms. The last two terms in the first line (i.e. $dSdt$ and $(dt)^2$) are ignored because they are small. However, unlike the usual first-order Taylor series approximation, Ito showed that the third term is not small. To see this, from Equation (12.6),

$$dS = \mu Sdt + \sigma Sdz$$

so that

$$\begin{aligned} (dS)^2 &= (\mu Sdt + \sigma Sdz)^2 \\ &= \mu^2 S^2 (dt)^2 + 2\mu \sigma S^2 dtdz + \sigma^2 S^2 (dz)^2 \\ &\approx 0 + 0 + \sigma^2 S^2 \varepsilon^2 dt \\ &= \sigma^2 S^2 \varepsilon^2 dt \\ &= \sigma^2 S^2 dt. \end{aligned} \qquad (12.10)$$

The third and fourth lines are obtained by ignoring higher order terms and using Equation (12.1), that is,

$$(dz)^2 = (\varepsilon \sqrt{dt})^2 = \varepsilon^2 dt.$$

The last line in Equation (12.10) uses the relation

$$\text{Var}(\varepsilon) = E[\varepsilon^2] - [E(\varepsilon)]^2 \quad \text{(by definition of variance)}$$
$$= 1. \quad \text{(since } \varepsilon \sim N(0, 1))$$

Hence,

$$E[\varepsilon^2] = 1 - [E(\varepsilon)]^2$$
$$= 1 - 0 \quad \text{(since } E(\varepsilon) = 0)$$
$$= 1.$$

Substituting for dS and $(dS)^2$ in Equation (12.9), we have

$$dF = \frac{\partial F}{\partial S}(\mu S dt + \sigma S dz) + \frac{\partial F}{\partial t} dt + \frac{1}{2} \frac{\partial^2 F}{\partial S^2}(\sigma^2 S^2 dt)$$

$$= \left[\mu S \frac{\partial F}{\partial S} + \frac{\partial F}{\partial t} + \frac{1}{2} \sigma^2 S^2 \frac{\partial^2 F}{\partial S^2} \right] dt + \sigma S \frac{\partial F}{\partial S} dz.$$

This completes the proof.

The next task is to build a portfolio P of x shares at $\$S$ per share and sale of one option at price F. Suppose we select $x = \partial F/\partial S$ so that the value of the portfolio is

$$P = S \frac{\partial F}{\partial S} - F. \tag{12.11}$$

Then the change in portfolio value in time dt is

$$dP = dS \frac{\partial F}{\partial S} - dF \tag{12.12}$$

since $\partial F/\partial S = x$ is a constant. If there are no arbitrage opportunities, this change is equal to the interest earned by the portfolio by investing in a risk-free asset at interest rate r, that is,

$$dP = rPdt.$$

Eliminating dP from the last two equations gives

$$dF = dS \frac{\partial F}{\partial S} - rPdt. \tag{12.13}$$

Equations (12.8) and (12.13) may be equated to give

$$dS\frac{\partial F}{\partial S} - rPdt = \left[\mu S\frac{\partial F}{\partial S} + \frac{\partial F}{\partial t} + \frac{1}{2}\sigma^2 S^2 \frac{\partial^2 F}{\partial S^2}\right]dt + \sigma S\frac{\partial F}{\partial S}dz,$$

that is,

$$dS\frac{\partial F}{\partial S} - \sigma S\frac{\partial F}{\partial S}dz = \left[\mu S\frac{\partial F}{\partial S} + \frac{\partial F}{\partial t} + \frac{1}{2}\sigma^2 S^2 \frac{\partial^2 F}{\partial S^2} + rP\right]dt. \qquad (12.14)$$

The left hand side may be simplified as

$$dS\frac{\partial F}{\partial S} - \sigma S\frac{\partial F}{\partial S}dz = (dS - \sigma Sdz)\frac{\partial F}{\partial S} = \mu Sdt\frac{\partial F}{\partial S}$$

using Equation (12.6). Hence, Equation (12.14) becomes

$$\mu Sdt\frac{\partial F}{\partial S} = \left[\mu S\frac{\partial F}{\partial S} + \frac{\partial F}{\partial t} + \frac{1}{2}\sigma^2 S^2 \frac{\partial^2 F}{\partial S^2} + rP\right]dt. \qquad (12.15)$$

This last equation may also be simplified as

$$0 = \frac{\partial F}{\partial t} + \frac{1}{2}\sigma^2 S^2 \frac{\partial^2 F}{\partial S^2} + rP. \qquad (12.16)$$

From Equation (12.11),

$$P = S\frac{\partial F}{\partial S} - F$$

so that

$$rP = rS\frac{\partial F}{\partial S} - rF. \qquad (12.17)$$

Substituting for rP in Equation (12.17) into Equation (12.16) gives

$$0 = \frac{\partial F}{\partial t} + rS\frac{\partial F}{\partial S} + \frac{1}{2}\sigma^2 S^2 \frac{\partial^2 F}{\partial S^2} - rF. \qquad (12.18)$$

Equation (12.18) is the Black-Scholes (1973) option pricing model. The difficulty now lies in solving it for the option price F for which closed form solutions are generally not available.

In the special case of a call exercised only at expiry (also called a European call), we let $F = C$ and the solution is

$$C = SN(d_1) - Ke^{-rT}N(d_2) \qquad (12.19)$$

where

C = price of option call;
S = current stock price;
K = strike price;

$N(x)$ = standard normal distribution function

$$= \frac{1}{\sqrt{2\pi}} \int_{-\infty}^{x} e^{-y^2/2} dy ; \qquad (12.20)$$

$$d_1 = \frac{\log(S/K) + (r + 0.5\sigma^2)T}{\sigma\sqrt{T}} ; \qquad (12.21)$$

$$d_2 = d_1 - \sigma\sqrt{T} ; \qquad (12.22)$$

σ = volatility;
T = time to expiry; and
r = risk-free interest rate.

Note that log(.) refers to natural logarithm (or log to base e), and it is useful to remember that log to the base 10 is rarely used in scientific work. The use of log(.) rather than ln(.) to refer to natural logarithm avoids the problem of "ln" being misinterpreted as "1" or "n". Usually, "n" is reserved for sample size. In this book, log(.) is used consistently to refer to natural logarithm.

To find the historical volatility of the price of a share, consider the discrete model

$$S_{t+1} = S_t u_t$$

where u is a random term and t denotes time. The *simple return* is defined as

$$r_S = (S_{t+1} - S_t)/S_t.$$

The *log return* is defined as

$$r_L = \log(S_{t+1}/S_t) = \log(u_t).$$

If $\log(u_t) \sim N(\mu, \sigma^2)$, then u_t is said to have a log-normal distribution. Empirically, the distribution of $\log(S_{t+1}/S_t)$ is close to log-normal for most share prices and this implies volatility (σ) may be estimated as the standard deviation of $\log(S_{t+1}/S_t)$. This is the rationale for using log returns rather than simple returns.

Hence, given annual stock prices $S_1,..., S_n$, it is a simple matter to compute the log returns

$$h_1 = \log(S_2/S_1)$$

$$...$$

$$h_{n-1} = \log(S_n/S_{n-1}).$$

Then an estimate of the variance of log returns is

$$\sigma^2 = \frac{\sum (h_i - m)^2}{n-2} \tag{12.23}$$

where m is the mean of log returns, that is, the mean of the hs. Taking the square root in (12.23) gives the historical volatility based on movements in past share prices. Note the denominator is $n - 2$ because there are only $n - 1$ terms.

If monthly rather than annual stock prices are available, the computed monthly volatility is annualized by multiplying it by $\sqrt{12}$. Similarly, if quarterly share prices are used, the quarterly volatility is annualized by multiplying it by $\sqrt{4}$. To see this, suppose we have quarterly stock prices $S_1, S_2,..., S_5$. Then

$$\log \frac{S_5}{S_1} = \log \left[\frac{S_5}{S_4} \frac{S_4}{S_3} \cdots \frac{S_2}{S_1} \right]$$

$$= \log \frac{S_5}{S_4} + \log \frac{S_4}{S_3} + \cdots + \log \frac{S_2}{S_1}. \tag{12.24}$$

If we let

$$h = \log(S_5/S_1),$$

the annual log return, and

$$H_4 = \log(S_5/S_4)$$

$$...$$

$$H_1 = \log(S_2/S_1),$$

the quarterly log returns, then from Equation (12.24),

$$h = H_4 + \cdots + H_1.$$

Hence, the annual volatility is given by

$$\sigma^2 = \text{var}(h) = \text{var}(H_4 + H_3 + H_2 + H_1)$$
$$= \text{var}(H_4) + \text{var}(H_3) + \text{var}(H_2) + \text{var}(H_1) \qquad \text{if } H_i, H_j \text{ are uncorrelated}$$
$$= 4\,\text{var}(H) \qquad \text{if } \text{var}(H_i) = \text{var}(H_j) = \text{var}(H)$$
$$= 4q^2$$

where q is the quarterly volatility. Hence,

$$\sigma = (\sqrt{4})q.$$

In other words, the quarterly volatility computed from quarterly share prices need to be multiplied by $\sqrt{4}$ to obtain the annualized volatility used in the Black-Scholes model.

Example

Let us return to the land option where the current land price is $100 per m^2 and the seller (land owner) offers the developer the option to buy the second site at $105 per m^2 to be exercised in two years' time. What is the value of this option?

Here,

$$S = 100; \qquad\qquad K = 105; \qquad\qquad r = \text{risk-free interest rate} = 0.04;$$
$$T = 2 \text{ years}; \qquad\qquad \sigma = 0.2.$$

If historic annual land prices are available, it is possible to estimate the volatility using Equation (12.23) and taking the square root. This highlights a difficulty with real options – unlike stock prices, data on land prices may not be readily available. From Equation (12.21),

$$d_1 = \frac{\log(S/K) + (r + 0.5\sigma^2)T}{\sigma\sqrt{T}} = \frac{\log(100/105) + [0.04 + 0.5(0.2^2)]2}{0.2\sqrt{2}}$$

$$= (-0.049 + 0.12)/0.283 = 0.251.$$

Hence, from Equation (12.22),

$$d_2 = d_1 - \sigma\sqrt{T} = 0.251 - 0.2\sqrt{2} = -0.032.$$

From the Appendix, the cumulative probabilities are

$$N(0.251) = 0.599; \text{ and}$$
$$N(-0.032) = 0.488.$$

Hence, the price of the call is

$$C = SN(d_1) - Ke^{-rT}N(d_2)$$
$$= 100(0.599) - 105e^{-0.04(2)}(0.488)$$
$$= 59.9 - 47.3 = \$12.60.$$

Recall that the land owner offers the option at a price of only $10 per m^2, below $12.60 per m^2. The developer should buy the underpriced option.

The literature on option pricing is still growing and only the main ideas have been sketched here. There are many qualitative introductions (e.g. Strong (2005)) as well as more quantitative texts (e.g. Hull (2003), Luenberger (1998), and Wilmott et al. (1995)).

12.12 Determinants of exchange rate

Since many infrastructure projects nowadays involve currency risks, some brief notes on the determinants of exchange rates are in order.

Historically, international trade was on dominated by silver coins. Gold was used more as a store of value. However, as trade volumes rose after the Industrial Revolution, England had a shortage of silver coins. Across the Atlantic, the US government incurred a large deficit to finance its War of Independence, pushing silver coins out of circulation. Thereafter, the *gold standard* slowly came into existence in the late 19th century, starting from Germany in 1871. Coins and notes bearing the king's face and minted at the King's mint (for a fee, of course) could no longer be taken at face value but were backed by gold. This created several problems for various parties.

One obvious problem was the incentive for the monarch to dilute the quantity of gold to finance costly wars and other luxurious expenditures, making it difficult to ascertain value. This debasing of the value of currency hindered trade by the raising transaction costs of ascertaining the purity of coins. Silver and copper were often used, and some coins were bimetallic.

A second problem was that gold discoveries would lead to over-minting and increased the money supply. This created high inflation, and countries with high inflation tended to have higher prices for their exports. In turn, high export prices reduced exports. Meanwhile, high domestic prices also encouraged imports at lower prices. The net effect of high inflation was to reduce exports and increase imports.

The third problem was that countries with persistent trade deficits would eventually run out of gold. If a country could no longer pay for its imports, it became protectionist. Once "beggar thy neighbor" protectionism spread, global trade would suffer.

Historically, it was the need of governments during World War I that marked the end of the gold standard. Without sufficient gold to print or mint currencies to finance the imperialistic wars, country after country began to leave the international gold standard. As protectionism spread, global trade slowed and contributed to the Great Depression of the 1930s.

After World War II, the US dollar served as the international currency under the *gold exchange standard* (note it differs from the pre-war gold standard). This Bretton Woods Agreement (1944) also included the setting up of two important international institutions. The International Monetary Fund was established to assist countries with persistent balance of payments difficulties with emergency loans. The World Bank was also set up to provide loans for post-war reconstruction of damaged infrastructure in many economies.

Like the outdated gold standard, the gold exchange standard was also backed by gold or, more precisely, the greenback was convertible to gold at a fixed exchange rate of US$35 per ounce. During the 1950s and 1960s, world trade flourished under the gold exchange standard and the World Bank became an important banker to many newly independent developing countries interested in accelerating national development.

The gold exchange system ended in the early 1970s after the US overprinted the greenback. Since the dollar was "as good as gold," greenbacks could be printed to finance her costly wars (e.g. Vietnam War) and American acquisition of assets overseas.

It was soon clear to other countries that the US did not have sufficient gold to convert greenbacks to gold on demand. When France pressured the US to convert French holdings of greenbacks (from the sale of French assets to the Americans) to gold, President Richard Nixon declared in 1971 that greenbacks were no longer convertible to gold. This ended the gold exchange standard.

In its place is the current *floating exchange rate* regime where currencies are no longer backed by gold. The value of a currency, or its exchange rate relative to another currency, is based on demand and supply. This is shown in Figure 12.8 for the price of the Australian dollar (A$) relative to the Singapore dollar (S$).

Price of A$ (= S$ to 1 A$)

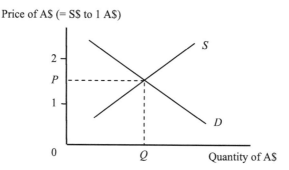

Figure 12.8 The exchange rate for A$.

On the demand side, Australian dollars are demanded by

- an importer of Australian products;
- a bank that has to pay interest on deposits that belong to Australians;
- a company that needs to pay dividends or profits to Australians;
- a government that gives financial aid to Australia; and
- a foreigner who invests in Australian assets (e.g. houses).

Conversely, the following people supply Australian dollars to the foreign exchange market:

- an exporter to Australia;
- a non-Australian who receives interest in A$;
- a non-Australian who receives dividends or profits from an Australian company;
- people who receive financial aid from the Australian government; and
- an Australian who invests in assets overseas.

The A$ demanded and supplied will "balance" in the foreign exchange market at the equilibrium exchange rate (Figure 12.8).

The above discussion is often written in technical jargon where exchange rate is determined by

- the balance on goods and services (i.e. exports less imports);
- net investment income (e.g. interest and dividends);
- net investment in assets; and
- government transfers.

Over the long term, a positive balance on goods and services will tend to strengthen the currency.

On net investment income, if a country has a high real interest rate relative to the rest of the world, it will attract short-term funds (or "hot money") into the country as deposits. To make these deposits, foreigners will need or "demand" A$ in the foreign exchange market and, if the supply is fixed, the A$ will strengthen. In particular, if the currency is weakening, real interest rates may need to rise to compensate investors from switching their deposits elsewhere. This explains why world interest rates in real terms tend to follow US interest rates.

On the speculative side, speculators may "attack" a currency if they feel it will weaken and this may destabilize the financial system. In the early 1990s, many "emerging economies" in Asia liberalized their financial and real estate markets, attracting substantial amounts of speculative money into the country. As stock and property prices rose, many construction projects were started using money borrowed from international markets and denominated in US dollars. Countries with currencies pegged to the US dollar saw their currencies appreciating in tandem with the greenback following the Clinton Administration's new policy of the early 1990s to reduce the large US budget deficit.

It soon became clear to some speculators (particularly hedge funds) that many Asian currencies were over-valued and short-selling occurred, beginning with the Thai Baht in 1997. Banks and large corporations panicked and sold their holdings of the Baht as well, making it impossible for central banks to defend the Baht after spending billions of dollars in frenzied buying of the currency. The currencies of Korea, Malaysia, Philippines, Taiwan, Hong Kong, and so on were similarly attacked within months of each other.

With massive falls in the value of domestic currencies, borrowers of funds denominated in US dollars went bankrupt as they could not service their loans. Many of these borrowers were governments, developers, contractors, and other businesses. Even local businesses that borrowed in local currencies were not spared; as foreign investors panicked and sold out, central banks hiked interest rates to stem the outflow of badly needed capital. With sky-high real interest rates, many businesses simply folded, unemployment rose sharply, and the property market collapsed.

Questions

1   Why may a project manager be interested in a forward or future contract?

2   Explain how an interest rate swap may be used to hedge the interest rate risk in a project.

3   Give examples of real options in each of the following cases:
     a) option to expand;
     b) option to defer;
     c) option to renew; and
     d) option to abandon.

4   Explain why it is more difficult to price a real option as compared to a financial option.

5   What determines the price of a call option?

6   Explain how a project with a negative net present value may still be financially viable if options are available.

7   A developer has acquired a piece of land at $100 per m^2 to build a facility. It has also been given an option to buy an adjacent site for $110 per m^2 in four years' time. If the risk-free interest rate is 3 per cent and volatility of land returns is 30 per cent, what is the fair value of the call option? [$24.57]

8   What are the limitations of the Black-Scholes option pricing model?

13

Agreements, Contracts, and Guarantees

13.1 Types of agreements, guarantees, and contracts

The main parties in a project finance structure (see Figure 8.1, reproduced here as Figure 13.1) are bound by certain key agreements and contracts. The general terms and conditions in these agreements, guarantees, and contracts are outlined in this chapter.

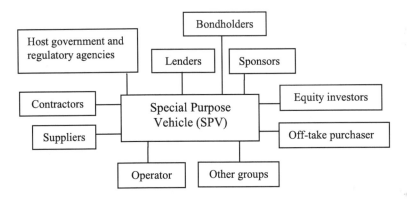

Figure 13.1 The structure of project finance.

13.2 Functions of contracts

Agreements, contracts, and guarantees are risk management instruments. As discussed in Chapter 9 (Section 9.6), contracts (and to a lesser extent agreements) are used to

- share or shift the price, output, and other risks;
- provide incentives for producing quality goods through proper requirements and specifications; and

- prevent hold-up or opportunistic behavior because of various types of asset specificities.

While they tend to reduce risks, contracts may also contain loopholes because of our cognitive limits (bounded rationality) in recognizing many types of contingencies and inevitable linguistic imprecision. Indeed, long and detailed contracts may contain more ambiguities and therefore increase rather than reduce the risk.

The most common causes of construction disputes and claims relate to (Netherthon, 1983; Lutz et al., 1990; Semple et al., 1994)

- deficiencies in contract documents;
- soil conditions;
- design deficiencies;
- defective specifications;
- scope creep; and
- scheduling problems.

These causes vary across projects. For instance, soil conditions are a major source of disputes in transit projects that involve extensive tunneling (Halligan et al., 1987). High quality contract documents and improved geotechnical information are required.

13.3 Remedies

If there is a breach of contract, compensation may be sought. Generally, parties try to negotiate privately among themselves for a win-win outcome. The authority to settle disputes is important; if it is too centralized, disputes may remain unresolved at the field staff level.

If private settlement is not possible, mediation by an independent third party, "project neutral," or Dispute Review Board may be required. However, the mediator's decision is not binding. If a binding decision is required, then arbitration is used. If all else fails, the parties go to court for expensive litigation.

"Partnering" is sometimes suggested as a means of resolving construction disputes. It is used more frequently in other industries where a long-term relation exists among customers, producers, and suppliers. "Partners" hope to seek win-win solutions, value long-term commercial relations, and cultivate trust and openness in resolving issues.

Generally, the contractor is required to keep working on the project during a dispute. In some cases, the practice is to resolve all disputes only upon project completion. This approach does not encourage prompt resolution, but it does allow more time and room for negotiation.

It is important to realize that the law does not provide remedies for all cases, and therefore does not necessarily cover all losses as well. Hence, one should not assume that claims against any wrongdoing would result in adequate legal compensation.

Claims may be categorized and made in respect of

- debt;
- breach of contract;
- tort;
- statute; or
- unjust enrichment.

Tort includes wrongs outside the categories of contract and statute, and common examples of such acts include nuisance, negligence, trespass, defamation, and deceit.

A claim under one category (e.g. contract) may fail but succeed in another (e.g. debt). For instance, a contractor who is not paid after completing a house may fail in his claim for damages if the client counter-sues for defective work. However, he may succeed in his claim for debt (Davenport, 1995).

13.4 Shareholders' Agreement

Generally, sponsors form an incorporated limited liability joint venture (JV) company, often in corporate form but in some cases, as limited partnerships.

The Special Purpose Vehicle (SPV) is a legal entity and owns all the project's tangible and intangible assets and is managed through its board of directors appointed by sponsors, passive investors (shareholders), and the government or its agency (as a shareholder).

The SPV is governed by a Constitution on matters such as

- the appointment of directors, number of directors, and limitation on powers of directors;
- restrictions on raising capital;
- introduction of additional parties; and
- pre-emptive rights in relation to transfer of ownership.

There is also a Shareholders' Agreement outlining

- the scope and structure of project;
- implementation schedule (e.g. start and end of JV);
- ownership interests;
- equity contributions;
- composition of management;
- roles and responsibilities;
- rules for voting and decision-making;
- rights and obligations including contingent financial support and disclosure of information, and reserved roles (such as if a sponsor is also appointed as the contractor without competitive bidding);
- procedures for feasibility studies;
- procedures for the appointment of consultants;
- management of contracts;

- mechanisms for resolving disputes; and
- distribution of profit.

The SPV is subject to auditing and reporting requirements as well as company taxation. For this reason, sponsors may not own the SPV directly but through an intermediary holding company located in a tax-favorable third country.

13.5 Implementation Agreement

An Implementation Agreement (IA) is signed among the relevant parties (e.g. sponsors and State agencies) to go ahead with the project, possibly conditional on the availability of public funds.

It spells out the goals of the project and the specific commitments of each party (see below). In addition, the parties undertake to make reasonable effort to consult and cooperate with each other through a Liaison Committee and agree on the means for resolving disputes.

The roles and responsibilities of each party generally cover

- land acquisition;
- provision of certain infrastructure;
- development of adjacent land; and
- ways of dealing with externalities and environmental problems.

Example

A mining firm may enter an IA with the Forestry Department and Wildlife Agency to implement a forestry and wildlife conservation plan for the area. The IA would cover

- technical assistance to be given by the agencies to the firm in the preparation of the conservation plan;
- the implementation process; and
- remedies if any party fails in its obligations (such as compensation, suspension, and revocation of license), and mechanisms for resolving disputes.

The implementation process includes start and end dates, funding, reporting, monitoring, and responses to environmental damage.

13.6 Loan Agreement

The project company (or a sponsor) and lenders enter into a Loan Agreement to provide debt financing for the project. The key provisions of the Loan Agreement include

- loan amount, fees, currency, term, rate of interest, debt service structure, grace period, and key dates;
- lender's right to terminate the agreement or accelerate repayment before maturity because of misrepresentation by the borrower;
- preconstruction affidavit to protect the lender against liens;
- lead manager or arranger (in a syndicated loan);
- collateral or security;
- guarantees (e.g. sovereign guarantee if a sub-sovereign government is also a borrower, or guarantees from a sponsor's parent organization);
- agreement to budget for debt service, including setting up a reserve fund for loan repayment;
- requirement to provide periodic financial data, often in the form of financial ratios such as ratio of current assets to current liabilities or a minimum working capital (i.e. current assets less current liabilities) to enable the lender to monitor the financial health of the borrower;
- disbursement procedure and payout schedule;
- penalties for late payment;
- priority of debt repayment with the loan as senior debt;
- restrictions on dividends and other distributions (e.g. profits), which can be an unpopular covenant;
- restrictions on further borrowings from other lenders to prevent excessive debt accumulation;
- restrictions on leasing and capital investment;
- obligation to maintain, repair, and insure assets;
- agreement not to sell a material portion of the assets (usually up to 20%) without the consent of the lender (contra disposal clause);
- agreement not to undertake illegal activities;
- restrictions on incurring contingent liability;
- maintenance of borrower's identity by not merging with another commercial entity without the lender's consent;
- triggering of default if there is any material adverse change in the financial health of the borrower as determined by the lender; and
- remedies in case of default including standby financing in terms of equity or subordinated loans to meet unexpected cost escalations or changes in market conditions.

While these covenants are intended to ensure that the borrower repays the loan, enforcing and monitoring them may be a different matter. For instance, financial ratios are merely rules of thumb on financial soundness. Further, there is a danger that the lender may impose unnecessary restrictions on the borrower and increase the cost of borrowing.

Project finance loans are often syndicated because of the risks and large sums involved. The usual process is for the borrower to identify his needs and then invite bids from banks to be the lead manager. The appointed lead manager will then prepare the information memorandum and circulate it to prospective lenders. It contains information on the lender, loan amount, basic loan terms, borrower's balance sheet and income statements, significant contracts, capital and expenditure forecasts, senior management, analysis of competitors, and country risk exposure analysis.

Once offers are received by the lead manager or arranger, a loan agreement is drawn up and the syndicate is formed. The lead manager and agent banks should exercise due diligence tackling with the thorny problem of projecting future cash flows. An exculpation clause is often inserted to exempt the lead manager from potential liability in the case of default. Whether such an escape clause is effective is still a matter of legal opinion, and the bank should conduct careful analysis, or due negligence, on its own.

Project loans may be made in single or multiple currencies (called a currency pool), and there are restrictions on the number of currencies. A single currency loan is more common, and some lenders (e.g. Asian Development Bank) are exploring the possibility of lending in local rather than hard currency. This is to overcome the mismatch between project revenue received in local currency and loan repayments in hard currency. Loans may be made directly to the SPV or, if the borrower is a State agency, to the government who, in turn, makes a sub-loan to the agency. There is potential for funds to be siphoned off in this roundabout way of financing a project.

There is a front-end origination fee of about 0.2 per cent of loan amount to compensate lenders for processing the loan. There is also a commitment fee of about 0.75 per cent of undisbursed balance. This fee is required to compensate lenders for setting aside the money irrespective of whether it is used. This charge is often viewed as an unnecessary cost and unpopular with borrowers. Clearly, sponsors can lower the commitment fee by

- phasing a development carefully so that the loan can be arranged in tranches and made available for disbursement only when necessary;
- borrowing for shorter periods; and
- borrowing in smaller amounts, resulting in lower undisbursed balances.

However, borrowers who borrowed in small amounts will have to undergo repetitive processing of loans, and more needs to be done by lenders to facilitate the loan approval process.

Indexed variable interest rates are common in project finance loans to shift interest rate and inflation risks to borrowers. Interest rates are often about 0.5–1 per cent above LIBOR. Increasingly, loan terms have become shorter (e.g. 5–7 years, down from 7–10 years) to increase lender's liquidity and because of the high risks of project failure despite the nexus of contracts, agreements, and credit enhancements. Projects are highly levered and susceptible to interest rate and other risks. There is also a grace period for loan repayment. Typically, it is about 3–5 years.

The term structure of interest rates for project finance loans tends to show a "hump" (Figure 13.2). That is, interest rate spread tends to rise initially and then fall for loans with longer maturity for up to 10 years and possibly longer (Kleimeier and Megginson, 2001). This implies that longer maturities in projects are not viewed as all that risky since a major part of project risks occur during the construction stages of a project.

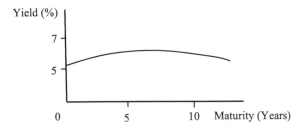

Figure 13.2 Term structure for project finance loans.

Kleimeier and Megginson applied ordinary least squares regression to a sample of 1,803 project finance loans from 1980–2000 (with an average size of US$177 m) and found that

$$Spread = 131 - 0.01S - 0.89M - 42.67G - 42.16C + 1.50c + 15.99A + e$$
$$(-1.31) \quad (-2.00) \quad (-11.27) \quad (-6.95) \quad (10.87) \quad (3.75)$$

Adjusted $R^2 = 0.17$.

where

Spread = loan spread above LIBOR in basis points;
    S = loan size in US$m;
    M = loan maturity in years;
    G = dummy variable = 1 if a loan has a third-party guarantee
                   = 0 otherwise;
    C = dummy variable = 1 if a loan is exposed to currency risk
                   = 0 otherwise;
    c = country risk rank where low-risk countries have lower ranks;
    A = dummy variable
         = 1 if the borrower is in an industry rich in collateralizable assets
         = 0 otherwise; and
    e = residual.

Although the model fit based on adjusted R^2 is low, most of the t statistics shown in brackets are significant at 0.05 level (i.e. greater than 1.96). The size of loan (S) does not affect loan spread above LIBOR, which is not unexpected. Large loans are likely to be syndicated and this lowers default risk for each bank arising from loan size.

Evidence for the hump in loan spread is given by the negative value of the estimated coefficient for loan maturity (−0.89). This indicates that loan spread does not increase positively in a linear fashion with maturity. As explained earlier, project risks need not rise with loan maturity; after a project has been constructed, the major risk associated with construction has been substantially lowered.

The negative coefficient for third-party guarantees (−42.67) is expected since such guarantees reduce the risk of default.

Interestingly, the coefficient for currency risk is negative (−42.16), that is, greater currency risk exposure reduces loan spread, which is "surprising." Kleimeier and Megginson suggested that lenders offer lower rates to borrowers who are willing to accept currency risk, though it is not clear why this would not be offset by greater default risk. One possibility is that the downside of currency risk may be minimal. Since many projects loans are denominated in US dollars, lenders expect the currency to appreciate, and therefore offer lower loan spreads. The other possibility is that exposure to currency risk has not been properly captured by the dummy variable.

The positive coefficient for country risk (1.50) is expected. Greater exposure to country risk increases the probability of default.

The positive coefficient for collateralizable assets (15.99) requires explanation. Kleimeier and Megginson offered two reasons. First, industries with collateralizable assets may be riskier, and borrowers therefore incur higher spreads. Second, these riskier projects tend to be funded using project finance, and therefore form part of the sample of 1,803 project loans in their study.

A caveat is in order. It has been noted that the R^2 of 0.17 is extremely low, which means the bulk of the variation in loan spread is still left unexplained by the model. Many project variables such as the procurement method, time of loan, experience of designers and contractors, and so on have been left out of the regression model.

13.7 Security Agreement

Since lending is non-resource or limited recourse, security is limited to

- the project assets,
- project rights (e.g. insurance claims or right to operate a facility),
- agreements,
- power to set tariffs to raise revenues,
- concessions, and
- cash flows.

The Security Agreement between the security trustee and SPV is intended to secure to lenders debt repayment. The trustee (such as a bank) administers the debt service and this includes setting up an escrow account with the bank to prevent the siphoning of project revenues for other purposes (Figure 12.3).

If a sub-sovereign government is also a sponsor (borrower), the Agreement may include revenue intercepts - the use of future central government transfer payments for debt service. A binding clause binds a subsequent government to continue with debt service.

Sometimes, interim financing is available to help the SPV tide over cash flow problems. The bank may offer a line of credit for the SPV to borrow up to a certain amount. For a fee of about 1–3 per cent, a standby lender may provide standby commitment to the SPV. If the SPV is unable to find a permanent lender, the standby lender will provide the permanent loan at a premium rate. If there is a shortfall between the construction and permanent loan, gap financing may also be arranged (Figure 13.4), which is similar to a bridging loan.

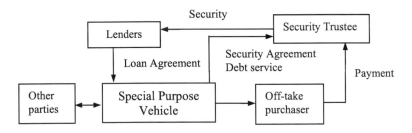

Figure 13.3 Roles of security trustee.

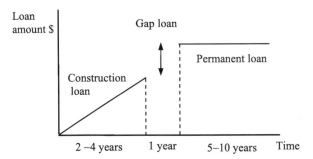

Figure 13.4 Gap financing.

13.8 Purchase Agreement

A long-term Purchase (or Off-Take) Agreement between the SPV and off-taker or purchaser of the output is usually imposed by lenders as a condition for the project loan to minimize market risk.

If the off-taker is a public authority such as an electricity board that buys electrical power from the SPV, payment can be problematic. In such cases, the government may be asked to guarantee payment.

There are many types of Purchase Agreements. In a take-or-pay agreement, the off-taker either purchases (takes) the product or pays the SPV in lieu of purchase. In a take-and-pay agreement, the off-taker pays for agreed quantities taken or minimum amounts over a specific period and price. The price may be fixed beforehand, indexed or based on prevailing market prices. Price floors and caps may also be applied in hedging agreements. If prices are fixed beforehand, the output is often sold at a discount to encourage sales and pre-development take-up as a requirement for the SPV to raise capital. Since long-term purchase agreements bind the purchaser and expose him to considerable price risk, large discounts may be expected.

For power projects, there may be a contract for difference where electricity is pre-sold by generators to retailers (e.g. utilities firms) for distribution to final consumers at a pre-determined or strike price. If the actual price differs from the strike price, either the generator or retailer will refund the difference depending on whether the actual price is above or below the strike price. Generally, the strike price is kept low so that electricity will be taken by the retailer for a number of years. If the actual price from the electricity pool is higher, the retailer will pay the generator the difference.

In a throughput or transportation Purchase Agreement, the off-taker or user of a pipeline agrees to use it to carry a minimum volume of fluid per unit time (e.g. monthly) at a minimum price. The off-taker has to pay for the pipeline service even if it does not transport the minimum volume.

13.9 Concession Agreement

A Concession Agreement between the SPV and public agency specifies the rules for the SPV to operate locally and the types and period of concession. For example, in a road project, the concession may specify that the firm finances, upgrades, builds, maintains, and operates a road for 20–30 years before transferring ownership to the government.

Additional rules may be specified, such as supervision of construction by a third party (e.g. independent consultant), periodic reporting, insurance, penalties for infringement, conditions for early termination of concession, as well as proper maintenance and audit of accounts.

Usually, a maximum toll is set and indexed for inflation. If usage falls below a specified minimum, the public authority may need to provide a grant to cover costs. Minimum service provisions are also specified as well as penalties for failure to provide adequate levels of service.

13.10 Supply Agreement

The Supply Agreement is a long-term agreement between the SPV and supplier of materials. The period of purchase, (indexed) price, and specifications are specified in the agreement to ensure adequate materials supply at pre-determined costs and quality.

To minimize currency risk, payment is usually denominated in the same currency as that of revenues.

13.11 Construction contract

As noted in Chapter 6, there are many procurement methods such as

- traditional lump-sum fixed contracts;
- design and build contracts;
- turnkey contracts; and
- management contracts.

Typically, standard forms of contracts are used to allocate risks and responsibilities, for example, the

- International Federation of Consulting Engineers (FIDIC) contract;
- Joint Contracts Tribunal (JCT) standard form of building contract;
- American Institute of Architects (AIA) contract for construction;
- Institute of Civil Engineers (ICE, UK) engineering and construction contract;
- Australian standard AS4000 general conditions of contract; and
- Singapore Public Sector Standard Conditions of Contract for Construction Works.

The general provisions of construction contracts include

- scope of work;
- commencement of works;
- project company risks and responsibilities;
- contract price;
- fluctuations clauses pertaining to changes in the price of materials, if agreed between the parties;
- claim procedure;
- progress payments;
- variations or change orders;
- supervision;
- right to accelerate works;
- definition of completion;
- force majeure;
- liquidated damages;
- suspension and termination;
- security; and
- mechanisms for resolution of disputes.

These general provisions do not preclude provisions found in the specific forms of contract listed above. Construction work may commence upon receipt of a notice to proceed (NTP) issued by the project company. If the SPV is unable to secure the financing, it has the right to terminate the construction contract after compensating the contractor.

The SPV assumes certain risks and responsibilities with respect to construction. These include

- ensuring site availability;
- ensuring access to utilities and payment to the contractor for new installations;
- incurring the cost of unforeseen subsoil conditions;
- obtaining the required permits;
- advance payment (e.g. 10 per cent of contract value, if any);
- ensuring third-party contracts are executed;
- accepting the risks for changes in law to pay for unanticipated fees and charges;
- providing fuel and other materials for plant testing; and
- incurring the cost of removal of hazardous waste.

Generally, the risk of differing subsoil conditions is borne by the SPV since it should not have bought the piece of land based on its consultant geotechnical report, that is, the SPV is best able to bear this risk. If the risk is shifted to the contractor, the latter faces an unacceptably high risk and requires a higher mark-up on cost. Even though he does not bear the risk, the contractor will also carry out his own independent soil test.

Advance payment may be made to the contractor prior to construction. This is followed by progress payment depending on certification of work done or agreed milestones provided claims for payment are made promptly.

Payments are also made for authorized variations or change orders (as distinct from free and informal work requests) due to

- design changes required by law;
- design changes initiated by the SPV;
- additional scope of work;
- design changes caused by unexpected events (e.g. changes in soil conditions);
- errors and omissions; or
- alternative designs suggested by the contractor, sometimes called value engineering but the term is more appropriately applied to steps undertaken to improve efficiency, that is, "add value" by reducing costs.

A change order is sometimes understood as a variation order. In other contexts, a change order is viewed as requiring more substantial changes than a variation order. However, it is difficult to draw the line, and the real issue is whether the contractor will be paid for carrying out the order, not whether the change is minimal or substantial. Claims against variation or change orders can be messy over where the impact of variation starts and ends, and rates for overheads and profit mark-up should be fixed beforehand to avoid disputes.

The SPV appoints a project manager to supervise the construction. Independent checkers may also be appointed to check structural designs and temporary works.

Projects that are substantially completed are initially accepted to commence operations after passing commissioning or user acceptance tests (UATs). Usually, non-critical parts such as landscaping may not be completed when a temporary occupation permit (TOP) is issued by the authority. Thereafter, there will be a defects liability period of about a year for the contractor to rectify all reported defects without additional cost.

Force majeure events are beyond the control of the contractor and could not have reasonably been anticipated. They include "acts of nature" (fire, earthquake, adverse weather, and flood), war, civil unrest, terrorism, and nationwide labor strikes. Not all acts of nature are unanticipated; if the site is prone to flooding, it is not a force majeure event.

Liquidated damage (LD) arises from late completion or failure of the installation to perform as specified. Delay LDs are calculated on a daily basis with a cap on the total sum. Performance LDs are computed based on projected loss in revenue or increase in operating cost as a result of the failure to perform to the required level.

The SPV has the right to suspend or terminate a contractor for poor performance.

The various types of bonds and insurance required have been discussed in Chapter 10.

13.12 Operation and maintenance contract

The Operation and Maintenance Contract between the SPV and facility operator specifies the responsibilities of the operator (including staffing), takeover procedures from the contractor, the period of operation, the maintenance and repairs required, and fees. The operator is required to secure the necessary permits to operate the plant.

13.13 Guarantees

A sovereign guarantee may be required in the following cases:

- the purchaser of the project output is a State agency and the sovereign is required to guarantee payment;
- the borrower is a State agency and the sovereign is required to guarantee its loan repayment; or
- when an international lending agency such as the World Bank requires a counter-guarantee.

In a counter-guarantee (Figure 13.5), the World Bank provides a partial credit guarantee (i.e. a guarantee for a portion of the loan) to commercial lenders for a fee. In turn, the Bank may require the sovereign to counter-guarantee repayment by the SPV, particularly if it is a State agency. The purpose of the World Bank guarantee is to encourage commercial lenders to lend to the SPV by reducing the risk of default.

The World Bank may also provide partial risk guarantees (for a fee) to commercial lenders against non-performance of sovereign contractual obligations or from force majeure events. Alternatively, lenders could purchase political risk insurance.

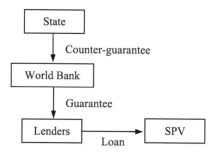

Figure 13.5 Counter-guarantee by a State.

Parent companies are usually asked by lenders to provide guarantees for their subsidiaries or associated companies. Generally, it will only guarantee up to the amount for which it is responsible. For instance, if it owns only 30 per cent of the SPV, then it is prepared to guarantee only up to 30 per cent of the latter's debt. Such guarantees are normally noted in the parent company's annual report as contingent liabilities. Since the SPV is a vehicle company, the lender may impose covenants on the guarantor.

In many instances, the SPV is a joint venture and each parent company will be asked to act as guarantor. If the partners are severally or individually liable, then they are liable for up to a certain portion of the loan in the event a guarantor is unable or unwilling to pay. However, if they are jointly and severally liable, each party may be liable to the entire amount of the debt if, for any reason, a guarantor is unable to fulfill its obligations. This is an unsatisfactory position a parent company should avoid.

Questions

1   Explain why claims against any wrongdoing may not result in adequate compensation.

2   Despite the nexus of contracts, agreements and guarantees, many projects still fail. Why?

3   Explain why the term structure of a project finance loan tends to show a hump.

4   If risks should be allocated to the party that is best able to handle it, explain why the World Bank is better able to bear political risks than commercial lenders.

5   In recent years, governments have been reluctant to counter-guarantee loans borrowed by State agencies. What are the implications?

6   What should a parent company consider in acting as a guarantor to a loan incurred by its subsidiary or associated company?

7   Explain why partnering has not reversed adversarial relations in the construction industry.

8   A major issue in construction disputes relates to costs. Yet, contractors are reluctant to release proprietary cost data, making it difficult for other parties to assess claims. How can this problem be mitigated?

Case Study I: Power Projects

14.1 Introduction

Chapters 14 to 17 contain case studies of various types of commercial and infrastructure projects. These case studies are developed from combinations of actual projects rather than just a single project to maximize learning outcomes. This approach is in line with the emphasis in this book on general principles or models rather than on specific projects.

From experience, case studies from single projects tend to be repetitive even though the cases are often representative. A major reason for this defect is that each case is presented in the same format using the project cycle approach. To overcome this problem, the cases in this book depart from some project case studies in a second way by focusing on different aspects of projects in each chapter. This approach avoids the repetition and provides the opportunity to discuss certain issues in depth.

The underlying thread in these case studies is to understand

- the specific markets (and institutional arrangements) of various types of projects;
- how the project is financed; and
- why a bundle of risk instruments in a complex project finance structure may fail and how such failures give rise to new risk management strategies.

The first case study concerns the financing of power plants. Power generation is a risky and capital-intensive business. Consequently, project financing is used extensively in this sector.

14.2 Types of power plants

The production of electrical power may be divided into three main sections, namely,

- the *generation* of power, often at a remote location to be close to water, wind, coal, oil or gas;

- the *transmission* of power from the power station to urban and rural communities over long distances using high-voltage, high-capacity transmission lines; and
- the transformation to lower voltage and *distribution* of power in populated areas, called the grid, to final consumers. This involves power retailing, that is, marketing, metering, and billing.

In the generation of electricity, wind power is seldom used on a large scale even though it has attracted considerable attention over the last two decades because of its cleaner technology and falling costs. However, even large-scale wind generators can only generate up to a few megawatts (say c MW), and

$$\text{Annual output} = c \times U \times 8,760 \text{ MWh}$$

where U is utilization rate (e.g. 0.3 or 30 per cent of the year when there is sufficient wind), $8,760 = 24$ hours $\times$ 365 days, and MWh is megawatt hour. In remote windy locations, wind generators have the advantage in supplying electricity to small communities.

Nuclear power plants are also less common because of safety concerns. Most communities are not keen on nuclear power plants since the Three Mile Island and Chernobyl disasters in 1979 and 1986 respectively. However, with high oil prices and greenhouse emissions from fossil fuels, some governments may again be keen on apparently cheaper nuclear power plants. Other concerns with nuclear power plants include the price and availability of uranium, and permanent storage of spent fuel.

Hydro-electric plants may require the damming of rivers with large differences in elevation to allow the natural flow of water to turn the turbine. In some cases where bank conditions and geology allow, it is cheaper to divert the water flow rather than dam the river. The latter may necessitate the resettlement of towns and villages due to flooding as well as the destruction of historic sites, forests, and wildlife.

Coal or diesel-fired plants use steam to turn the turbine. More recent cogeneration plants use a combined cycle where gas is used to turn the turbine and the heat generated is then used to make steam to generate additional electricity. Hence, the efficiency of cogeneration plants can rise to as high as 80 per cent compared to 60 per cent for single cycle plants. However, the initial cost of a cogeneration plant is correspondingly higher.

14.3 Feasibility of power plants

On the cost side, the cost of a power plant includes

- land cost;
- construction cost;
- financing cost; and
- operating and maintenance costs.

The cost of land is not easy to determine given that these plants are often situated in remote areas without recent land sales to compare unit land values following a change of use. However, power projects are commonly procured on build-operate-transfer or build-operate-own basis so that the land is likely to have been provided by the government. Hence, the land price may be taken out of commercial consideration.

The construction cost at the feasibility stage is often estimated using the unit method by comparing its capacity (in megawatt, or MW) with comparable plants. The cost is usually indexed for inflation as well as regional or international variations in material and labor prices. Typically, indexation is not done for equipment prices because capital is mobile.

In more precise estimates, the general work breakdown structure of a power plant is first carried out. This breakdown may include the building structures, turbine, connecting road, bridges, earth transport, spillway, river diversion, and relocation of pipeline. In some cases, quantities for each component are estimated by engineers and given to shortlisted contractors in the Request For Proposal (RFP) for pricing and bidding. In other cases, quantities are estimated but not given to contractors to encourage them to optimize their design.

The financing cost is not estimated separately once the total cost of construction is known. Rather, it is incorporated into the cash flow analysis as discussed in Chapter 6 (see below).

The annual operating and maintenance costs may be estimated with reasonable accuracy as a percentage of construction cost. These simple cost ratios obtained from comparable plants are quite stable.

On the demand side, revenue from the plant is not too difficult to predict because electricity is a necessity and consumption patterns do not deviate substantially from year to year. In other words, demand for electricity is price and income inelastic in the short run and slightly more elastic in the long run as reactions to price changes take effect.

However, empirical estimates of the elasticities of electricity demand for businesses and residential users are fraught with technical difficulties such as nonlinear pricing (or multi-part tariffs), rendering them unreliable because of the wide variance (Taylor, 1975; Bohi, 1981; Filippini, 1995).

Finally, the existing number of firms and households that will be served by the plant may be reasonably estimated. A growth factor is then applied to forecast annual electricity demand.

As discussed in Chapter 6, these revenues and costs are then used as inputs into a pro forma cash flow analysis together with

- depreciation based on an estimated project life;
- interest expenses;
- tax considerations; and
- loan repayments.

The net present value, project internal rate of return, and equity internal rate of return are then computed. The reader is reminded that many short-cuts such as the payback period or return on investment are available but effort should be made at this stage to reduce the financial uncertainties by executing a more elaborate feasibility study using proper discounting procedures.

14.4 Traditional financing arrangements

Traditionally, the generation, transmission, and distribution of power were owned by the government through State-owned or government-linked enterprises. It was believed that the generation of electricity is a natural monopoly because of falling long-run average cost (i.e. increasing returns to scale), making it difficult for more than one firm to operate in the industry. The firm with the lowest cost will undercut other producers.

The transmission and distribution of power were also thought to be natural monopolies but for a different reason to that of generation. It would be uneconomic for two firms to duplicate the transmission infrastructure. Similarly, duplicating the grid was thought to be wasteful.

To gain and sustain popular political support among businesses, farmers, and workers, electricity prices were deliberate kept low in many countries. However, there are several serious consequences, namely,

- State-owned enterprises were not run on commercial basis and lacked the profit motive;
- the subsidy strained the State budget;
- supply exceeds demand because of over-consumption, leading to chronic shortages and rationing;
- electricity prices were fixed nationally to despite regional cost differences; and
- cross-subsidy among different types of users (e.g. businesses, farms, and residential users) raises the question of equity.

This state of affairs, which continued to exist in many parts of the world up to the 1960s, was unsatisfactory.

14.5 Regulation of electricity prices

The road towards privatization started in the 1960s and 1970s with the price regulation of power generation. A government or quasi-government entity was allowed to produce electricity on commercial basis subject to price regulation. However, the transmission and distribution of power continued to be viewed as State monopolies to avoid duplicating costly infrastructure.

How can the price of electricity be regulated to prevent the monopolist from exploiting consumers?

Unlike many commodities, electricity is difficult (costly) to store, and supply matches demand at any time by altering generator frequency. At low capacities, the supply curve (S) is almost horizontal (see Figure 14.1). However, at maximum capacity (Q^*), electricity cannot be increased in the short run and the supply curve is nearly vertical. Supply cannot be increased by storage, inventory, or transfer from another source.

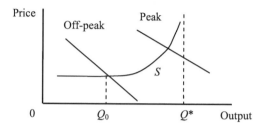

Figure 14.1 Peak and off-peak prices.

The demand for electricity (D) is highly seasonal, such as between summer and winter months. There are also intra-day and intra-week fluctuations between peak and off-peak demand. Thus, as shown in Figure 14.1, prices fluctuate sharply between the daily peak and off-peak demand, or between summer and winter months.

Regulators also need to consider that the monopolist faces a declining long-run average cost (AC) curve (Figure 14.2). For the firm, profit is given by

$$\pi = R(Q) - C(Q)$$

where R is revenue, C is cost, and Q is output. Both revenue and cost are functions of output.

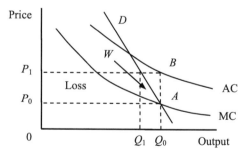

Figure 14.2 Pricing electricity.

The profit-maximizing output is obtained by solving

$$d\pi/dQ = dR/dQ - dC/dQ = 0.$$

Hence,

$$dR/dQ = dC/dQ.$$

i.e.

Marginal revenue = Marginal cost, or
MR = MC.

If the electricity price is set to marginal cost (i.e. P_0) to achieve social efficiency, the monopolist incurs a loss given by the rectangle P_0P_1BA. This loss needs to be covered by a State subsidy.

Alternatively, if the price is set to average cost (i.e. P_1), the monopolist breaks even but the output is socially inefficient. This is because, for outputs between Q_0 and Q_1, the marginal social benefit represented by the demand curve exceeds the marginal social cost. Hence, there is a deadweight loss (W) to society.

In practice, regulators tend to set the rate of return (s) on the firm's investment (K) as

$$sK = pQ - \text{Expenses} \tag{14.1}$$

Typically, s is fixed at a real rate of 10% and K is based on the historic value of investment. Operating, maintenance, and interest expenses are then estimated. Hence, given output Q, it is possible to set a regulated price p using Equation (14.1).

Apart from problems regulators have in accessing the firm's data, the above so-called "rate of return" regulation may result in

- too much capital investment if the cost of borrowing is less than s (Averch and Johnson, 1962); and
- insufficient incentive for the firm to reduce cost.

Consequently, performance standards, reviews, and benchmarking were used to make the monopolist more efficient. To further encourage the firm to reduce its cost, it was allowed to retain part of the earnings that result from cost-cutting measures.

14.6 Independent power producers

By the end of the 1970s, the view that power generation is a natural monopoly and therefore the producer needs to be regulated on pricing to earn a normal return (i.e. based on long-run average cost) began to change.

Importantly, the expectation that newer and more efficient turbines would result in falling long-run average cost did not materialize. In contrast, the optimal scale of generators actually declined, making it possible for more than one firm to operate in the power generation industry. In addition, transmission losses were reduced, and producers in one region could compete and supply electricity in another region.

These two reasons set the scene for deregulation of the power generation industry in the 1980s. It became possible to unbundle the three primary functions of generation, transmission and distribution, and as we have seen in Chapter 3, a vertically integrated firm is also doing too much and not focusing on its core competences.

Only transmission is now regarded as a natural monopoly and, to prevent duplication of the transmission infrastructure, several independent power producers (IPPs) would be charged similar prices to transmit power. Often, the transmission and distribution are still owned by the State, provincial government, or local enterprises.

Since the domestic capital market was weak, many IPPs were either foreign sponsors or joint ventures between foreign sponsors and State-owned enterprises. One such configuration is shown in Figure 14.3. The State-owned enterprise is still dominant through its subsidiaries as sponsor, off-take producer, fuel supplier, and operator. Its debt obligation is guaranteed by a State bank.

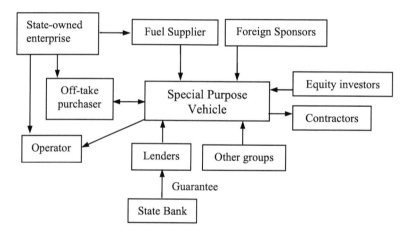

Figure 14.3 Example of project financing structure for power projects.

14.7 Risks in power projects

From the perspective of IPPs, the specific risks in power projects include

- access to transmission lines and distribution, which needs to be specified in the Implementation Agreement;
- the availability of long-term power purchase agreements (PPAs) with a single electricity buyer (such as the local electricity board) where payment could be guaranteed by the central government to reduce market (demand) risk;

- setting tariffs on terms similar to Equation (14.1) and indexed to a hard currency (e.g. US dollar), inflation, and fuel prices. In some cases, the tariff adjustment was based on agreed indexes or formulas. In other cases, approval from the appropriate ministry was required.

The configuration in Figure 14.3 contains other risks. The glaring weakness is the continual large involvement of the State-owned enterprise in multiple roles, resulting in problems such as

- conflicts of interest;
- weak management;
- overstaffing;
- accounting irregularities;
- complex decision-making and approval processes;
- poor quality of construction because contracts were awarded to related subcontractors; and
- supply of inferior fuel.

On the positive side, political risks were reduced if the State-owned enterprise owned a substantial share of the SPV. However, this is partially offset by political interference in managing and running the SPV.

In some configurations, the cash-strapped State-owned enterprise may transfer some of its existing power plants to the SPV and use them as collateral to raise its share of equity through an initial public offer. In many developing countries, these plants are generally not viewed by potential investors as high quality assets. The plants may be obsolete or technologically inferior. Another possibility is to issue power revenue bonds guaranteed by the State-owned enterprise or other sponsors, but problem of asset quality remains.

The State Bank guarantee was cold comfort if the State ran into serious financial problems. This occurred during the 1997–8 East Asian economic crisis. As the local currency was devalued, the State-linked off-take purchaser could not honor its obligation given the low electricity prices it charged to final consumers to garner political support. Hence, attempts were made to renegotiate power purchase agreements.

The same problem occurred in the Enron Dabhol project in the Indian state of Maharashtra. In 2001, the local off-take purchaser, the Maharashtra State Electricity Board (MSEB), was short of money and the State government had to provide emergency funds to help partially pay the then Enron-controlled IPP, Dabhol Power Corporation (DPC). This effort was made to avert another massive blackout if creditors forced DPC into liquidation. However, the State government complained that DPC's electricity prices were too high and sought to renegotiate the disputed PPA (which was signed in 1995) where MSEB was contracted to purchase all the power produced by DPC irrespective of demand.

Many onerous PPAs were signed secretly and lacked transparency. It was a recipe for corruption and bribery. In other cases, these long-term contracts were signed in anticipation of rising energy prices following the oil shocks of the 1970s and 1980s. When energy prices did not rise substantially in real terms in the 1990s, many purchasers were short of funds.

In recent years, international institutions such as the World Bank and Asian Development Bank have provided partial risk guarantees to commercial lenders against political risks. The Banks then passed such risks to private insurers or required counter-guarantees from governments. These guarantees covered the government's non-compliance with respect to

- breach of contract;
- convertibility and availability of foreign exchange;
- changes in law;
- force majeure; and
- expropriation of assets.

As we have seen, such guarantees tend to lose force if the local currency is substantially devalued.

14.8 Market pricing

During the 1997–8 Asian economic crisis, many investors withdrew their funds and invested in independent power plants in the US. The rationale was that nuclear power plants would shut because of safety concerns or play a smaller role and, unlike the over-supply of independent power generators in many developing countries, there would be a shortage of conventional power plants.

In addition, the new gas-fired generators were highly efficient. This raised the prospect of greater profitability and the possibility of opening the US power generation industry to more players.

As early as 1978, the Public Utility Regulatory Policy Act (PURPA) required utilities to buy power from qualified independent power producers. Not surprising, given the lack of transparency and opportunities for corruption, many of the initial long-term PPAs were signed with rather high prices. When the playing field was widened in the 1990s, these PPAs were increasing replaced by new dynamic wholesale electricity markets. The wholesale price of electricity was determined round the clock either hourly or half-hourly. But while the utilities bought electricity at dynamic wholesale prices, the retail price for electricity these utilities charged to consumers was fixed by regulation.

This new arrangement gave rise to three adverse consequences. First, consumers had little incentive to alter their demand for electricity during peak and off-peak periods because they were charged a fixed rate. Indeed, consumers merely flicked the switch when electricity was required, and could not observe intra-day prices.

Second, during peak periods, the wholesale price of electricity could soar well above the wholesale price (see Figure 14.1), forcing many utilities to file for bankruptcy. As a result, price or revenue caps are now common in the wholesale market.

Third, the fear of being left with unpaid bills prompted independent power producers to curtail supply simultaneously and the system shut down during 2000–1, creating a massive blackout (Borenstein, 2002).

Shortly after, the State of California had to step in to sign long-term PPAs with power producers to stabilize electricity prices, but these rates were 20 per cent higher than prices before the crisis, and among the highest in the US. There were also allegations of independent power producers forming a cartel to restrict supply to jack up wholesale prices and, like the problems in the developing countries, the long-term PPAs were overpriced.

The Californian episode may be seen as a failure to coordinate the deregulation process than a failure of deregulation. If wholesale electricity prices are determined hourly, then it does not make sense to fix retail prices way below wholesale peak-load prices.

After the crisis, the State of California has taken several steps to mitigate the risk of a shut down. These include

- higher retail prices;
- purchase of electricity from other states, which requires improving the existing transmission infrastructure;
- the use of alternative sources of power; and
- building more plants to increase supply and weaken the market power of power producers.

In the end, the lesson is clear: using the market involves certain risks. The issue is not market or regulation but the design of appropriate institutions for markets to function properly.

Questions

1 Explain why the traditional method of providing electricity to consumers is unsatisfactory.

2 Why is the provision of electricity susceptible to market failure?

3 Explain why the multiple roles played by a party in structuring project finance may be a liability.

4 Discuss the possibility of an SPV issuing bonds to finance the building of a new power plant.

5 Explain why the Californian electricity crisis of 2000–1 occurred.

6 Despite deregulation, electricity prices have generally not fallen substantially (if at all) nor converged across countries. Why?

7 There are many creative ways of issuing bonds to finance infrastructure. In October 2006, as part of its effort to raise $1 bn, Keppel Corporation issued 5-year notes with 100 per cent principal protection at maturity but coupons are linked to the performance of an equally weighted basket of six commodities, namely, crude oil, nickel zinc, copper, gold, and silver. The minimum

investment is S$5,000 or US$5,000. According to the distributors (i.e. banks and brokers) of this issue, investors can potentially earn between 6–11 per cent of coupon per year depending on the movements in the commodity indexes. Are such notes worth investing in, and why?

15

Case Study II: Airport Projects

15.1 Introduction

In the previous chapter, we considered project financing of power projects and regulatory problems. It was shown that long-term power purchase agreements could break down because of uncertainties over the short-term and long-term pricing of electricity. Even when dynamic hourly pricing was used, the system could fail if there was a difference between unregulated wholesale and fixed retail electricity prices.

This chapter deals with airport development projects. Based on purely commercial considerations, many airport projects are not viable. However, if benefit-cost analysis is used, such projects may be socially viable.

Another difference between power and airport projects is that there is no long-term purchase agreement of airport services in the latter. In other words, airports are built on speculative or anticipated demand, and possibly under intense competition.

15.2 Monopoly of airline services

Like electricity generation, the provision of airline services was initially thought to be a natural monopoly. The barrier to entry arising from the large capital required to purchase or lease planes was considered to be too high for more than a few players.

In addition, national airlines were more than just commercial carriers. Many of these airlines were substantially owned by the government for strategic as well as commercial reasons. To ensure the success of these airlines, many domestic and international routes were tightly regulated in terms of landing rights, gate availability, capacity, and frequency.

Prior to the 1980s, most international airlines operated with limited landing rights, gates, or frequencies of operations. Gates owned by airlines, extracted through concessions from airport operators, severely restricted access by other airlines.

On the domestic front, competition was minimal. The field was opened to only a few operators. For most part, international airlines could not operate domestic routes in other countries.

With limited competition in both international and domestic routes, prices for airline services were mostly unregulated and consequently high. Differential pricing was the norm. Travelers in different classes were charged different prices to reflect differing ability to pay, costs, service, and market power.

15.3 Deregulation of airline services

In the 1980s, with modern access to global funds, raising the capital to purchase planes was no longer seen as all that problematic. This potentially allowed for greater competition among existing airlines and new airline operators to enter the field.

In addition, the idea that the average cost of airline services decreases with output was not supported by empirical evidence. Early studies of airline services prior to the construction of larger and more fuel-efficient long-haul planes showed constant economies of scale (White, 1979). Hence, the larger airlines did not have significant cost advantage over smaller ones.

There were also inefficiencies from welfare loss due to monopolistic pricing by airlines. In Figure 15.1, S is the horizontal supply curve, D is the demand curve, P is the price of airline services (such as a one-way ticket for a flight between two cities), and Q is the quantity of airline services demanded (number of passenger trips). Prior to deregulation, the price is p^*.

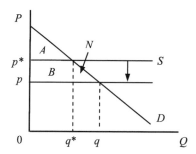

Figure 15.1 Welfare loss from monopoly.

After deregulation, competition forces the price down to p and the quantity demanded rises from q^* to q. If the regulated price is fixed at p^*, then quantity demanded is q^*.

Since the demand curve may be interpreted as the marginal social benefit curve and the supply curve is the marginal social cost curve, the marginal social benefit exceeds the marginal social cost for outputs between q^* and q. Existing consumers benefit from the additional consumers' surplus given by the area of the rectangle B but this is offset by an equivalent loss of revenue suffered by the airlines. The net benefit to new consumers is the area of triangle N. Hence, the net welfare gain to society from deregulation of airline services is N. This is the case for deregulation of airline services.

15.4 Privatization of airports

The privatization of airports is a different matter. Traditionally, airports were built, operated, and subsidized by governments. Many airports are still built using this system of financing.

In the 1980s, some governments were unable or reluctant to continue subsidizing airport infrastructure because of fiscal stress or were convinced by the "tide of market liberalization" that the days of bureaucratic and inefficient central planning as well as militant trade unionism were over. The turning point was 1981 when President Ronald Reagan, who just took office, fired striking air traffic controllers.

In 1987, several airports in the UK were privatized, and this soon spread to many parts of the world. The privatization of airports paved the way for building or upgrading airports using project finance.

Often, airport charges were not regulated after privatization. Instead, informal monitoring is used to ensure owners and operators provide an acceptable level of service.

15.5 Hub and spoke networks

As domestic routes became deregulated, airlines increasingly began to operate their flights using regional hub and spoke networks. This reduced the number of direct flights but increased load factors and the number of connecting flights for most passengers.

It then became important to regional governments whether their airports, if they have one, functioned as hubs or spokes. These decisions were inevitably made on political and economic grounds. For regional airports that began to function as hubs, domestic air traffic began to rise rapidly with increasing inter-connectivity. This often necessitated the upgrading of its airport infrastructure to improve connectivity, safety, and level of maintenance. Although most international air routes were still tightly regulated to protect the interests of national carriers despite many bilateral and multilateral "open skies" agreements, many regional airports were upgraded in the 1980s to attract international long-haul flights that used larger aircrafts.

In the 1990s, budget or "no frills" airlines began to operate a separate system of low-cost direct flights with reasonable frequencies instead of the hub and spoke system of traditional carriers that left passengers "stuck" with more transit points at crowded airports. This alternative arrangement provided a new lease of life to smaller regional airports. Threatened by the new and smaller competitors on an untapped market segment, some traditional airlines set up new subsidiaries (e.g. SilkAir and Tiger Airways of Singapore Airlines) to compete head-on with the new budget airlines.

Nonetheless, fierce competition, excess capacity, the successful entry of some budget airlines, security problems, and other external shocks (e.g. oil shocks, SARS, and bird flu) contributed to the bankruptcies of many small and large carriers. With aggressive undercutting of fares, airlines undertook measures such as closing down less lucrative routes, improving productivity, sharing excess capacity, and mergers.

In September 2001, the highly cyclical industry was hit by terrorism, resulting in further structural transformations. The already-crowded hubs were further slowed by tighter security checks on travelers unhappy with long delays, rising airport taxes, baggage restrictions, and fuel surcharges. Not surprisingly, airport inefficiencies came under scrutiny (Gillen and Lall, 1997; Hooper and Hensher, 1997).

15.6 Feasibility of airports

A feasibility study of airports should take into consideration the above points on the historical development of the airline industry.

The feasibility of upgrading airports that operate as hubs is usually not as complex as the upgrading of spokes because of difficulties in forecasting passenger and cargo demand.

From Figure 15.1, the net gain to society each year is

$$N = \tfrac{1}{2}(p^* - p)(q - q^*) = \tfrac{1}{2}\Delta p \Delta q. \tag{15.1}$$

One can then apply a reasonable growth rate to determine the net gain for each year (N_t) and compute the project IRR (k) using

$$0 = -C_0 + \frac{N_1}{1+k} + \cdots + \frac{N_n}{(1+k)^n}. \tag{15.2}$$

Here, C_0 is the estimated initial project cost, and n is the lifetime of the project. For instance, if the estimated price fall is $20 per trip and the number of new passenger-trips is one million, the net benefit for the first year of operations is

$$N_1 = \tfrac{1}{2} \Delta p \Delta q = \tfrac{1}{2}(20)(1,000,000) = \$10 \text{ m.}$$

If a growth rate of 5 per cent is applied, then

$$N_2 = N_1(1 + 0.05),$$
$$N_3 = N_2(1 + 0.05),$$

and so on. If the airport costs $100 m to upgrade, then

$$0 = -100 + \frac{N_1}{1+k} + \cdots + \frac{N_n}{(1+k)^n}. \tag{15.3}$$

If n is known, it is possible to compute k.

15.7 Air cargo and other services

If the airport handles cargo as well, the net benefit from cargo handling (M_t) must be added to that from air travel to determine the net benefit for each year. Hence, Equation (15.3) becomes

$$0 = -100 + \frac{M_1 + N_1}{1+k} + \cdots + \frac{M_n + N_n}{(1+k)^n}.$$

In additional, an airport provides other services such as aircraft maintenance, fuel, and in-flight catering.

15.8 Annual benefits and costs

In summary, the annual benefits and costs of an airport upgrading project typically consist of the following:

Initial cost:

Land acquisition
Upgrading of runway
Upgrading of terminals
Upgrading of control tower
New equipment
Information system
Fuel hydrant system
Construction of jet bridges
Civil works
Environmental costs
Drainage

Annual benefits:

Benefit to additional passengers
Additional cargo handling
Landing and gate fees
Retail rents
Catering rents
Hanger charges
Car parking fees

Annual costs:

Loan repayment
Operating costs
Maintenance and repairs

These benefits and costs may be estimated and used to compute the equity IRR using the method discussed in Chapter 6. As before, the environmental costs are not estimated and left as a political decision. There are also substantial indirect employment benefits in aircraft maintenance, in-flight catering, tourism, retailing, and so on.

It is relatively easy to over-estimate revenues and under-estimate the initial and operating costs.

15.9 Politics of airport projects

The politics of airport projects depends on the constellation of forces. The major stakeholders may include

- sponsors;
- politicians and bureaucrats;
- lenders;
- large landowners;
- businesses;
- workers;
- speculators;
- squatters;
- mass media;
- small absentee landlords; and
- renters.

The possible problems include

- protests against destruction of environment;
- complaints about aircraft noise;
- uncooperative State agencies, resulting in delays;
- backlog of acquisition cases because of inadequate compensation or disputes over land ownership;
- poor integration with existing infrastructure;
- failure to build related infrastructure such as railways;
- financing problems;
- disputes between lenders and sponsors over managing of the project;
- procurement problems;
- problems with "anchor" airlines that use the airport as base and are undergoing restructuring;
- sale of "cheap" land to foreigners;
- allegations that the project is over-ambitious and a waste of public money; and
- escalation in house prices if the supply of local housing is inelastic.

There are possible solutions to these problems. For a start, the positive side of the airport project should be made known to the public to build a more balanced picture. The local community stands to benefit from new construction contracts, expanding tourism, better infrastructure, and job creation. Rather than a drain on public coffers, the additional taxes collected from passengers, airlines, businesses, and property owners from improved land values could be used to fund other community projects.

The destruction of the environment in airport projects is usually less serious than in power projects. Complaints about aircraft noise are often dealt with by imposing restrictions on operating hours.

Normally, airport projects will need to be managed by a high-level committee to overcome bureaucratic delays. This committee is also in charge of integrating the new project with the existing and supporting infrastructure, in acquiring the land, and in building alternative housing and premises for displaced residents and businesses.

15.10 Project finance structure

As noted earlier, airports are financed differently using a mixture of public and private funds. The discussion in this section uses a project finance structure for a typical airport upgrading project.

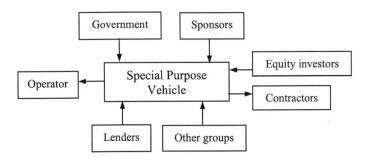

Figure 15.2 Project financing structure for airport projects.

The government may play the role of an equity investor and regulator. Usually, it will provide a long-term concession (e.g. 30 years) to the SPV under a build-operate-transfer project. To recoup its equity contribution or fund other upgrading projects, the government is likely to levy an airport improvement fee on each passenger or raise airport tax.

In return for giving the concession, the Civil Aviation Authority sets a maximum cap on the airport charges the SPV can levy on airlines. This is to ensure that the airport remains competitive. In the case of Sangster International Airport in Jamaica, the SPV has to pay a concessionaire fee to the Airport Authority of Jamaica based on

- passenger and cargo loads;
- 45 per cent of any gross revenue above the forecast amount; and
- half of any return in excess of 25 per cent.

Typically, the SPV is a consortium of a few sponsors with experience in either airport construction or operations. For example, in the Jamaican airport project (2001–8 (estimated)), the SPV comprises a transport firm, two contractors, and an experienced airport operator. The sponsors play multiple roles; in this case, one of the sponsors is also the EPC contractor, and another sponsor will operate the airport.

Thailand's Suvanabhumi Airport was funded under a different structure with substantial State involvement. Planning started as early as 1960 under the master plan for the Bangkok metropolitan area.

The airport site, Nong Ngu Hao (Cobra Swamp), is located at about 30 km east of Bangkok. In anticipation of the possible hike in land values arising from future economic growth and land speculation, the government purchased 3,100 ha of the land in 1973. The sequence of events leading to the opening of the airport in September 2006 is as follows:

- 1973 - Project was shelved when the military government was overthrown by a student uprising.
- 1978 - Project was reviewed by a consulting firm but again it was shelved.
- 1991 - Congestion at the Don Muang airport prompted the government to proceed with new airport under the charge of the Airport Authority of Thailand (AAT), the state-owned airport operator.
- 1993 - Master plan for the airport was awarded to foreign consultants and completed.
- 1994 - Contracts for ground improvement and flood control were awarded.
- 1994 - Design contract was awarded to international consultants after a major competition, but the glass-clad design was criticized as lacking in Thai architecture and unsuitable in Thailand's hot climate.
- 1996 - Relocation of 8,000 squatters. The shelving and revival of the project had encouraged residents to ask for higher compensation based on current market value, not values determined in 1973.
- 1996 - Change of government halted the project.
- 1996 - New government proposed building the new airport at Bang Pu but the idea was unpopular.
- 1996 - Formation of New Bangkok International Airport (NBIA) to revive the project but it was short-lived.
- 1997 - East Asian financial crisis and replacement of Thai government following a massive devaluation of the Baht.
- 1997 - The new government was committed to build the airport but had no money to fund the estimated US$3.7 billion bill.
- 1998 - Substantial modifications to the design to reduce the cost of construction.
- 1998 - Request by the government to its main lender, the Japan Bank for International Cooperation (JBIC), for additional loans.
- 2000 - Recovery from financial crisis.

- 2001 - First pile was placed without JBIC's loan approval, which upset JBIC. Prime Minister Thaksin Shinawatra, who just took office, said that Thais were ready to help themselves if JBIC did not approve the loan for a project that included Japanese contractors.
- 2002 - Laying of foundation stone.
- 2002–4 - Construction of the project with intermittent delays over the lengthy process of selecting contractors, use of costly imported materials, alleged irregularities over the tendering process, concerns over the weak roof design, alleged corruption, fire, differences between sponsors and architects, and poor integration with surrounding infrastructure.
- 2004 - Plans to integrate the new airport with road and rail links.
- 2005 - NBIA filed a lawsuit against a Bangkok newspaper for alleging large cracks in the runway. The report was immediately retracted.
- 2005 - JBIC approved additional US$300 m loan for the project.
- 2005 - Replacement of NBIA by Airports Thailand PLC.
- 2005 - Thaksin camped at the construction site to drive the slow-moving project. Critics argued he was merely looking for a showcase project to boost his chances of reelection.
- 2006 - 2,500 families that were relocated lodged a protest demanding that the promise of employment in the new airport given to them be kept.
- 2006 - Replacement of Thaksin government by a military coup because of graft allegations and his plans to reshuffle the military.
- 2006 - Opening of new airport in September 2006.
- 2007 - Cracks at runway and subsidence at many air gates raised safety concerns and the government has decided to reopen Don Muang Airport.

The planning and construction of Suvanabhumi Airport took 46 years, much longer than the smaller-scale Sangster International Airport.

Questions

1 Explain why the private profit calculus of airport operators tends to under-estimate the economic value of airports.

2 China's Zhuhai Airport, located about 70km west of Hong Kong, was built in 1995 at a cost of US$900 m. In 2006, the Zhuhai Airport Group (ZAG) entered a joint venture with the Hong Kong Airport Authority (HKAA) to form Zhuhai-Hong Kong Airport Management Company Limited to manage the airport for 20 years. ZAG holds a 45 per cent equity stake, and the remainder is held by HKAA. What are the major risks for HKAA?

3 The annual capacities of recently built or upgraded regional airports are given below:

- Thailand's Suvanabhumi airport (100 million passengers);
- Singapore's Changi airport (60 million passengers, after Terminal 3 is completed in 2008);

- Kuala Lumpur International Airport (40 million passengers, after Terminal 2 is completed in 2008); and
- Hong Kong International Airport (87 million passengers when fully developed).

These airports are currently operating below capacity. Explain why overbuilding may occur.

4   Identify the advantages and disadvantages when sponsors play multiple roles in airport projects.

16

Case Study III: Office Projects

16.1 Introduction

In the last two case studies, we saw the importance of mitigating market risks. In the case of power projects, long-term purchase agreements were used, and these agreements were so important for project success that some deals were apparently signed secretly and later created public uproar. However, even these purchase agreements could not withstand the pressure from a financial crisis where cash-poor governments sought to renegotiate terms.

In the case of the California electricity market, problems with long-term purchase agreements both in the US and other countries led to the use of dynamic market bidding on an hourly basis in the deregulated wholesale market. However, the fixed retail prices led to a system shutdown as utilities lost large sums of money during periods of peak demand when prices spiked. Suppliers were unable to respond to peak demand in the short run without escalating costs. Since the crisis, retail prices have been lifted to stabilize the system, and consumers pay higher prices.

For airports, some carriers build their own terminals in their "home base" or domestic airports but the predominant mode is the sharing of terminals. Airlines pay airport operators aeronautical fees for runway and terminals services and non-aeronautical fees for other types of services. An airline that uses an airport frequently is likely to enter into a long-term purchase agreement with the airport operator. Other long-term purchase agreements may include the signing of office leases and other terminal facilities.

In the case of commercial offices, long-term purchase agreements are relatively rare. Most offices are leased for only short periods such as for an initial three years and renewable for the same period. This makes office development projects risky, and specific risk management strategies are required. Some of these strategies are discussed in this chapter.

16.2 Dynamics of office markets

The dynamics of office markets differ from that of many other markets in several key areas, and these are outlined below.

Heterogeneity

Offices are not homogeneous units. They differ in location, accessibility, floor area, design, and other characteristics. Consequently, there is no ideal standard "unit" of office space.

In most aggregate (or macro) studies of office markets, heterogeneity is assumed away so that office space is measured in terms of the number of square feet or square meters without regard to quality.

Obviously, a prime office in a posh area is quite different from a run-down office unit. For this reason, office brokers tend to segment the office market into different grades. However, in most econometric studies of the office market, segmentation is ignored.

Durability

Office buildings are durable and last for many years. This means that single-period demand-supply analysis found in standard economics textbooks is insufficient. Instead, the two-period stock-flow model (Rosen, 1984; Hekman, 1985; McDonald, 2002) is usually used.

The office stock is the *existing* amount of rentable or sellable space in square meters or square feet. At any time t, the office stock (or equivalently the stock supply S_t) is relatively fixed because office production requires several years before a building is completed. Hence, as shown in the left panel in Figure 16.1, the stock supply curve (S) is vertical.

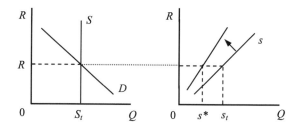

Figure 16.1 Stock-flow model of the office market.

The demand (D) for office space at time t is given by

$$D_t = a + bE_t + cR_t + e_t \tag{16.1}$$

where a is a constant or intercept term, b and c are coefficients or parameters to be estimated, E is employment in the office sector (e.g. finance, insurance, real estate, legal, and so on), R is effective real rent (nominal rent less rent-free periods divided by rate of inflation), and e is the error term, assumed to be normally distributed with zero mean and constant variance, that is, $e \sim N(0, \sigma^2)$.

Sometimes, E_t/E_{t-1} is added on the right hand side of the equation to capture the effect of a change in office employment but its effect is likely to be marginal. In some models, E is replaced by the rate of economic growth. Data on effective rent is based on a rent index. Quarterly data may be used to estimate the coefficients using ordinary least squares regression and if these data are not available, half-yearly or annual data are used.

In the left panel in Figure 16.1, the intersection of the demand and supply curves gives the equilibrium office rent R or, more precisely rent at time t, or, R_t. In some models, the office market is assumed to be in short-run disequilibrium so that rents adjust in response to vacancy movements. That is,

$$\Delta R_t/R_{t-1} = f + g(V_t - V^*) + v_t \tag{16.2}$$

where $\Delta R_t = R_t - R_{t-1}$ is the change in rent, f is a constant or intercept, g is a coefficient to be estimated, V is vacancy rate, V^* is the natural vacancy rate, and v is another error term.

The natural vacancy rate is independent of time (i.e. a constant) and may be estimated from a graphical plot of historical vacancy rates (Figure 16.2). It is usually assumed to be about 7%. It is the rate of vacancy caused by normal "frictional" adjustment of office space, a concept similar to the natural rate of unemployment to take into account inevitable frictional (rather than structural) adjustments when people are looking for work or changing jobs.

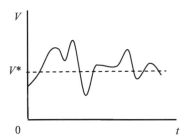

Figure 16.2 The natural vacancy rate.

Equation (16.2) states that rents will adjust if the actual vacancy rate deviates from the natural vacancy rate, which is intuitively reasonable.

The stock of office space at time t is given by the identity

$$S_t = (1 - d)S_{t-1} + s_{t-k} \tag{16.3}$$

where d is the straight line average depreciation rate for office buildings (e.g. $d =$ 0.01 if buildings last 100 years on average), S_{t-1} is previous office stock, and s_{t-k} is office building *started* (or commenced) k periods earlier. If annual data are used, the value of k is about three or four, that is, it takes three to four years for an office development to complete.

Equation (16.3) states that the current office stock is the sum of the previous stock (adjusted for depreciation) plus new completions. The latter depends on office developments started k periods earlier. In turn, office starts are given by the estimating equation

$$s_t = h + mR_t + nV_t + qLC_t + rI_t + \mathbf{p}'\mathbf{T}_t + u_t \tag{16.4}$$

where h is a constant, m, n, q, and r are coefficients to be estimated, R is effective rent as before, V is vacancy rate, LC is land and construction cost, I is interest rate, $\mathbf{T}$ is a vector of tax considerations so that $\mathbf{p}$ is a vector of coefficients and $\mathbf{p}'$ is its transpose vector, and u is another error term. Since data on LC (particularly land prices) are often not available, LC is often dropped from the model or replaced by an index of construction wages or an index of the cost of selected building materials (e.g. steel and concrete). If there are no major tax changes during the study period, $\mathbf{T}$ is also dropped and office starts are determined by only a few key variables (rent, vacancies, and interest rates).

Equation (16.4) states that developers start office projects based on current rents, vacancy rates, land and construction costs, borrowing costs, and tax considerations.

A simplified version of the stock-flow dynamics is given in Figure 16.1 where the intersection of demand and supply curves on the stock side (left panel) determines office rent R and, based on this rent, developers start to build s_t units of office space (right panel). In this simple version, all variables except rent in Equation (16.4) are held constant. This is merely a diagrammatic device to help us conceptualize what is going on, that is, the demand and supply of existing office stock determines the flow of new space that will come on-stream a few years later when buildings under construction are completed.

In reality, other variables cannot be "held constant" but this additional complexity does not present any serious problem in conceptualizing the adjustment process. For example, if interest rates and construction costs rise, then the cost of office development will rise, and the flow supply curve (right panel) will shift upwards as indicated by the arrow. The quantity of office starts will fall from s_t to s^*. Conversely, if construction costs and interest rates fall, then office development costs will also fall and the flow supply curve will shift to the right (not shown in the right panel). Developers will build more office space because, at the current rent level, it is profitable to do so.

In some models, the expected rent R_t^e is used in Equation (16.4) instead of the current rent since rental income is based on the market rent developers *expect* to rent out upon completion of the building, not at the point when they decide to build offices. Alternatively, it can be argued that expected rents are based on current rents or are predicted from current rents, that is,

$$R_t^e = \beta + \lambda R_t \qquad (16.5)$$

where β and λ are parameters. A further justification for using current rent is that pre-leasing is common in office developments and such rents are based on current levels.

Sometimes, the term ϕR_{t-1} is added to the right hand side of Equation (16.5) to capture the possibility that expectations may be adaptive, that is, they depend on the recent past (rent at $t-1$ here, or one period earlier). But it is seldom necessary to add this term. The current rent (R_t) is a better indicator on future movements in rents. It is interesting that, in economics and finance, expectations about the future influence current choices. If developers expect future rents to fall, they will scale down on development. However, if every developer scales down the workload, then land prices will fall because there are fewer bidders, and profitability rises because of falling costs. Hence, up to a point, it becomes profitable to start projects again.

The short-run equilibrium rent at time t is found by solving Equations (16.1), (16.3) and (16.4) for R_t, that is, by equating D_t to S_t, we have

$$R_t = \alpha_1 + \alpha_2 E_t + \alpha_3 S_{t-1} + \alpha_4 V_{t-k} + \alpha_4 R_{t-k} + \alpha_5 LC_{t-k} + \alpha_6 I_{t-k} + \mathbf{p}' \mathbf{T}_{t-k} + w_t \quad (16.6)$$

where the αs are coefficients to be estimated and w_t is another error term. This is the estimating equation for studying and predicting movements in office rents. If desired, Equation (16.2) may then be used to study the short-run dynamics of rent adjustment. It is evident from Equation (16.6) that if office projects are large, construction will take a longer time and k will be large. Yet, if k is large, the predictive power of the equation falls. It is unlikely that current rents are affected by distant events.

The above stock-flow model of the office market is a gross simplification of reality. It is a crude attempt to take into consideration the durability of office buildings. It reminds us to consider employment, the existing stock of office space, vacancies, previous rents, land costs, construction costs, interest rates, and taxes in understanding and projecting future rents. Mechanical predictions from historical data should be augmented with judgment.

Lumpiness

Another characteristic of office space is that its supply is "lumpy." That is, a large-scale office project adds a substantial amount of space when it is completed. When a few projects are completed at about the same time, there can be a substantial oversupply of office space. Rent predictions that were made when projects started a few years earlier would have gone wrong.

Apart from lumpiness, there are other reasons why over-building of office space occurs, namely,

- over-optimistic or naïve projections of office demand based on historical trends (or poor data) that do not take into account structural changes;

- impact of external or policy shocks;
- long planning times required for office development; and
- slow adjustment of office space to over-supply because vacancies cannot be cleared easily be changing the use of the building or exporting surplus office space.

Can governments do something about over-building? Looking at the causes of over-building above, the answer is yes, but the issue is by how much. Governments can

- provide better and timely data on office rents, starts, and completions for the private sector to plan;
- cushion the impact of external or policy shocks; and
- reduce planning and construction times by reducing approval delays.

However, there are problems with using the "visible hands" of governments. Bureaucrats and politicians may be out of touch and do not have better information. They may be incompetent, or do not understand the full consequences of policy actions.

16.3 Feasibility of office projects

The feasibility of office projects may be determined using the approach in Chapter 6 where NPV and project (or equity) IRR are computed. For instance, if equity IRR (q) is required, then

$$0 = -E_0 + \frac{F_1}{1+q} + \frac{F_2}{(1+q)^2} + \cdots + \frac{F_n}{(1+q)^n} \tag{16.7}$$

where E_0 is initial equity, Fs are cash flows, and n is the project life. If project IRR (k) is required, then

$$0 = -C_0 + \frac{N_1}{1+k} + \frac{N_2}{(1+k)^2} + \cdots + \frac{N_n}{(1+k)^n} \tag{16.8}$$

where C_0 is the initial cost of the project, the Ns are net operating incomes, and n is project life.

At the feasibility stage, the initial cost of an office development is determined using the unit method (e.g. per square meter). As the design progresses, the functional areas and building elements are worked out, measured, and then priced by contractors. The various procurement systems are discussed in Chapter 6. In many office developments, the traditional Design-Bid-Build method or the Design and Build delivery system is used.

As noted in the previous section, a key uncertainty is forecasts of cyclical office demand and rents. This means that the cash flows or net operating incomes in Equations (16.7) or (16.8) are only estimates.

16.4 Project finance structure

The project finance structure for office development differs from that of power and airport projects. As shown in Figure 16.3, there is often a syndicate of lenders who provide a short-term construction (and land) loan to the SPV *after* a permanent loan has been secured. The permanent lender (e.g. an insurance company or real estate investment trust) "takes out" the construction loan from the construction lender upon completion of construction.

The permanent lender is willing to hold the office asset for the long haul because

- it can rely on periodic office rents to match its cash flows;
- while office yields (returns) may or may not be attractive, there is potential capital gain at the end of the holding period when the property is sold;
- the project has been completed so that construction risk has been eliminated;
- market risk is mitigated by requiring the SPV to pre-sell at least 60 per cent of the sellable floor area; and
- the SPV may securitize the property asset (see below).

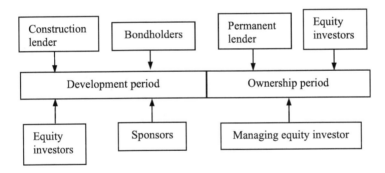

Figure 16.3 Possible financial arrangement for an office project.

Once the facility is built, one of the sponsors may become the managing equity investor who manages the facility in terms of marketing, collection of rents, maintenance, and repairs for a fee. For securitization to take place, the office development is likely to be held by the SPV during the first three years of operations to demonstrate its profitability and as a statutory requirement before its initial public offer to potential investors.

16.5 Risk management

Office development is risky because of volatile office demand and other risks. Several risk management strategies are outlined below.

Location

It is obvious that the project needs to be in the right location so that vacancy rates are relatively lower.

In neoclassical urban land use theory (e.g. Alonso, 1964), land users trade off transport costs with space. Users who incur small transport costs, such as residential users who make only a few trips a day to work in the city center or visit a neighborhood center, prefer to pay less per unit of urban land and live in the suburbs.

A large retail store or office block that needs to attract a large number of shoppers or office workers and hence incurs hefty overall transport costs will need to locate at a central location. By similar argument, a factory needs to be close to large pools of workers or major transport routes. It will tend to locate near ports or highway interchanges.

After buying a piece of relatively expensive land in a central location, developers substitute capital for land and build at higher densities.

These market outcomes are altered by planning controls over land uses and densities. In many cities, such controls tend to follow market outcomes to reduce over transport costs. For instance, it does not make economic sense to zone valuable land at low densities. If this is done, asset values and property taxes will fall.

Urban land use is also affected by history and durability of buildings. It is not uncommon to find old buildings (e.g. religious buildings, government buildings, and palaces) in the city center. These buildings are there for historical and architectural reasons.

Urban space is also a contested terrain. Local residents may protest against urban renewal, new highways, or the siting of "undesirable" land uses in residential areas.

The choice of intra-city location is also partly influenced by cultural preferences so that many large cities look like an ethnic mosaic with the usual Chinatowns and Little Indias. Gans (1962) called these people the "urban villagers" who prefer to live with "people like us."

The question whether politics, town planning, economics (i.e. land markets), history or culture is more important in shaping city form is difficult to answer. Politics and economics will feature highly. If it is a capitalist city, then most people would argue that capitalists often decide.

Over time, the city grows outwards because transport costs tend to fall or advances in communications technologies allow people to communicate relatively cheaply over longer distances. As a city grows outwards, suburban centers tend to form to cater to the growing suburban population as well as tap pools of workers to do back office work. High-tech science and industrial parks may exist alongside research universities.

A key question for an office developer is whether to locate the development in the city center or suburban center. If the rate of profit is equalized, the developer should theoretically be indifferent. However, once risks are considered, most developers tend to locate their offices in the city center. Despite the decentralization of city population towards the suburbs, many city centers are growing as well for precisely this reason: that it is less risky for the developer to locate in a central place.

Mixed commercial development

Since office development is cyclical, a central location by itself may not be insufficient to mitigate cyclical risk. Hence, some developers opt for mixed developments of compatible uses such as office space, convention halls, and retail spaces.

The mix must be organized so that the land uses are complementary. During weekdays, there are many office users who value the opportunity to shop nearby. The weekend crowds are attracted by conventions and shops.

Building convention halls is a risky business because of cyclical demand and oversupply of halls. Many halls are under-utilized during weekdays, and are not fully booked during weekends. With internet conferencing, firms are sending fewer employees to trade shows and conferences, and prefer to send them only to the larger ones. In the US, many optimistic projections of convention demand were based on naïve historical projections and off the mark (Sanders, 2002).

Unfortunately, there is no reliable method of forecasting the demand for convention halls. Many conventions tend to be one-off events and relatively few large conventions take place regularly at the same location. Further, changing internet technologies is a major threat to the convention business.

On the supply side, many cities built convention halls to attract a slice of the national or international convention business. It has considerable multiplier effects on airlines, land transport, hotels, shops, and the construction industry. Hence, convention centers may be built on subsidized loans or land to revitalize declining city center or boost city economic growth.

16.6 The case of Suntec City

Suntec City is a large mixed commercial development in Singapore. Its genesis may be traced to the way Singapore developed.

After World War II, Singapore fought for political independence and achieved self-government in 1959. In 1963, it joined the Federation of Malaya to form Malaysia but separated two years later because of irreconcilable differences. The government of Singapore then embarked on an extensive industrialization and modernization program.

The industrialization program of the 1960s and early 1970s was based on low-cost manufacturing to tap on Singapore's cheap labor. The modernization program included providing mass public housing, education, and health care as well as reinvigorating Singapore as a regional tourism, shipping, aviation, and financial hub.

By the late 1960s, urban renewal of the inner city to rid squatters was in full swing and, after a decade of growth amid the turbulent stagflation of the 1970s, it was decided that Singapore should also become a convention city to support the manufacturing and service sectors as the "twin engines" of its "second industrial revolution."

In the early 1980s, the world economy went into another of major recession not long after the second oil shock of 1979. In the US, interest rates were raised substantially by the Federal Reserve to boot out inflation. By 1984, the "twin engines" were faltering amid a global recession.

The then Prime Minister of Singapore, Mr Lee Kuan Yew, invited a group of Hong Kong tycoons (including Li Ka Shing) over to discuss ways to revive the sluggish economy. At that time, the tycoons were looking for a large project to earn stable incomes as part of a diversification strategy, well ahead of the inevitable return of Hong Kong to China in 1997. Consequently, in 1986, a company was incorporated in Singapore with an authorized capital of $100 m as a vehicle to realize this vision.

The next year, the Singapore government identified a large 11.7 ha site to build its first world class convention hall. The location was ideal as it was in the heart of the inner city, within walking distance to many hotels and the recently completed mass rapid transit (MRT), and just 25 minutes from Changi Airport. The tycoons formed Suntec City Development Private Limited (SCD) and won the land tender with a low bid of $200.9 m because of the property glut. The building cost was initially estimated at $810 m, and the total project cost was therefore slightly over $1 bn.

This huge sum was financed through debt and equity. Given the regional networks of the different tycoons and the project's solid fundamentals, debt could be raised without much fuss. About a dozen banks (including several local banks) were willing to lend $850 m to SCD. The sponsors, all with deep pockets, forked out $430 m in equity.

In 1989, plans for Suntec City were drawn up. The initial scope of work comprised

- a luxury hotel;
- four 45-storey office blocks (222,000 m^2);
- a large shopping mall (45,000 m^2); and
- Singapore's first purpose-built international convention and exhibition center (60,000 m^2).

The following consultants were appointed at various phases of the project:

Principal consultants

Design consultant
Project architect
Project manager
Civil, structural, and traffic engineers

Mechanical and electrical engineers
Quantity Surveyor
Accredited structural checker

Specialist consultants

Acoustics
Cladding
Façade maintenance
Fountain
Interior design
Kitchen
Landscaping
Land survey
Lighting
Retail design
Graphics
Theatre

The project was to develop in phases as follows:

Phase	Items
I	18-storey luxury hotel, one tower block, convention hall
II	2 tower blocks
III	2 tower blocks

In the same year, the $72.6 m piling contract was awarded, and financed using equity.

Interestingly, the developments in Suntec City were not pre-sold because of poor market timing. However, the tycoons were confident that the property market would pick up and spaces could be sold or rented at much higher prices than the depressed levels of 1989.

Soon after, it was realized that the project cost of $1 bn was a gross under-estimation because of major design changes and rising labor and materials costs as Singapore began to crawl out of recession. The new project cost was re-estimated at $1.5 bn, a 50 per cent increase from the 1988 estimate of about $1 bn. This escalation in cost required a few changes. Although raising the money was not viewed as a serious problem, SCD took a different path. It dropped plans to build the luxury hotel partly because of the glut and partly to save costs. The proposed hotel was replaced by an 18-storey office block.

In 1990, the substructure contract of $200 m was awarded after piling was completed. Shortly after, a new estimate put the project bill at $2.2 bn, or an increase of 47 per cent over the 1989 estimate of $1.5 bn, just before the award of the main contract of $1.03 bn. Between 1992 and 1993, construction proceeded at a fast pace after the award of the main contract and many smaller contracts (e.g.

cladding, air-conditioning, escalators, fountain, lighting, interior design, underpass, electrical equipment, and seating).

In 1994, SCD requested for additional funding, and lenders raised the loan amount from $850 m to $1.4 bn. However, the equity from sponsors was also increased from $430 m to $600 m.

The next year, near the height of the East Asian economic boom, the first office tower was completely sold floor by floor mainly to business associates and shareholders for $750 m. This gave the project a tremendous boost of confidence and dramatically eased the tight cash flow. It is a matter of speculation whether the sponsors had foresight or were sheer lucky.

The convention hall was officially opened in 1995 by none other than the Senior Lee who had invited the tycoons to Singapore in the first place a decade earlier.

In 1996, at the height of the East Asian economic "miracle," SCD had little trouble selling the next office tower, also floor by floor, for around 10 per cent more than the $750 m it pocketed a year earlier. Suntec City was also the venue for the inaugural Word Trade Organization Ministerial Conference.

Just when everything was going well for the tycoons, the East Asian financial crisis struck in July 1997, leaving SCD with three unsold office towers, a large retail mall, and a convention hall. Almost immediately, prime office rents fell steadily from $100 per m^2 to about $40 per m^2 per month by early 2004. When rents fell, assets prices fell as well to maintain yields (i.e. initial yield = annual rent/asset price). As sales fell sharply, the remaining unsold offices were leased. However, given the soft office market, rents were low, and vacancies were well above 10 per cent.

Fortunately, proceeds from the sale of the first two office towers were sufficient to reduce a substantial part of the debt. Hence, SCD was able to withstand another major blow in 2003 when the Severe Acute Respiratory Syndrome (SARS) spread to many Asian countries and slowed the short economic recovery from the financial crisis.

Like many large commercial developers, the tycoons were waiting for the right time to cash out. In 2004, German insurance giant Ergo offered to buy most of the remaining "profitable" assets in Suntec City. However, SCD was also looking at the alternative of securitizing the assets under Suntec Real Estate Investment Trust (Suntec REIT). This route was slower; it required the sponsors to show that rents were sufficiently high over three years before it could initiate a public offer of shares in the REIT.

In the end, the neater private purchase was rejected in favor of a public listing with a higher price tag. In December 2004, 722 million shares of Suntec REIT's initial public offer (IPO) were launched at $1 per share with an expected yield of 6 per cent. Revenue would be generated from its holdings of Suntec City's retail and office assets. The convention hall was not included in the REIT to maintain the target yield.

In all, Suntec REIT paid SDC $2.107 bn for the offices and retail mall comprising $1.9 bn at the onset and $0.207 m in deferred payments. Of the $1.9 bn direct payment, $1.335 bn was in cash and the rest were paid through a 43.9 stake in the REIT (565m shares). Part of the cash would be raised from the IPO, and the rest from borrowings by the REIT.

In 2006, Suntec City was the venue for the Annual Meetings of the Boards of Governors of the International Monetary Fund and the World Bank Group. Its retail outlets continue to do well, office rents have been rising steadily since the IPO in 2004. Two new train stations will be close by when the new Circle Line opens in 2010.

Questions

1   What are the market risks in a large-scale office development?

2   What are the strengths and weaknesses of the stock-flow office model?

3   Explain why office cycles tend to be more volatile than the general business cycle.

4   What were the major risks in the Suntec City project, and how were they managed?

17

Case Study IV: Chemical Storage Projects

17.1 Introduction

Chemical storage projects present a different set of challenges. The oil and gas as well as the petrochemical industries can lose or profit substantially from market turbulence arising from events such as an energy or geopolitical crisis. Poor coordination among producers can lead to oversupply, and acute shortages jack up prices and profits.

Capital investments are very high in such industries, of the order of millions or billions of dollars. Hence, there are relatively fewer large players, the use of project finance is common, and many cities compete to build petrochemical complexes because of the substantial investments and multiplier effects on other sectors.

Operational risks are also high. Safety and reliability are important in chemical storage, and the installation must be protected from fire, mischief, chemical leakage, lightning, floods, and acts of terrorism.

17.2 Organization of petrochemical complex

Petrochemicals are derived from two feed stocks, namely,

- natural gas liquids (e.g. ethane, propane, and butane) obtained from natural gas processing plants; and
- naphtha and gas oil from oil refineries.

The core of the complex is an upstream billion-dollar cracker and derivatives plant that produce feed stocks for downstream facilities (Figure 17.1).

Natural gas liquids are "cracked" at high temperatures to yield the basic petrochemical building blocks of ethylene and related products. Similarly, naphtha or gas oil yields are also "cracked" to produce ethylene and other products.

These basic building blocks of petrochemicals are then used as feed stocks by downstream plants to form secondary petrochemicals, chemical products, or synthetic resins for use in industrial and consumer products.

A large petrochemical complex requires the following services to function properly:

- access to large tracts of land to build the plants and store the feed stocks and petrochemicals;
- access to competitively priced feed stocks;
- access to funds;
- first-class logistics and support services; and
- first-class road, rail, and shipping infrastructure.

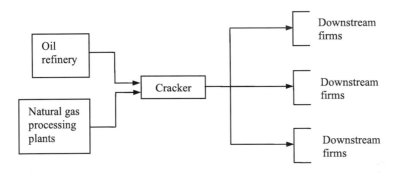

Figure 17.1 Basic organization of petrochemical complex.

For such large expensive projects, strong government backing is necessary to regulate the complex, provide tax incentives and infrastructure, invest as co-sponsor, and so on.

In an "advanced" petrochemical complex, there are significant long-term collaborations among upstream and downstream firms to integrate the entire process and sharing of common resources. In simpler complexes, there is usually a single ownership or limited collaborative and integrative arrangements among firms.

17.3 Shanghai Chemical Industrial Park

This brief description of Shanghai Chemical Industrial park (SCIP) is intended to provide some sense of scale to the above sketch of the framework of a petrochemical complex. More detailed information on SCIP may be found in the SCIP website.

The idea of developing SCIP was etched in China's 10th Five-Year Plan (2001–5) to build petrochemical parks in strategic locations in China. SCIP is one of four petrochemical bases that serve the Shanghai area. Another 16 petrochemical complexes are spread all over China to satisfy the huge demand for petrochemical products.

The developer of the park is the State-controlled Shanghai Chemical Industrial Park Development Company (SCIPDC). The park is administered by the SCIP

Administration Committee (SCIPAC). SCIPDC is an equity partner in some of the firms that operate in the park. Its simplified organization structure is given in Figure 17.2.

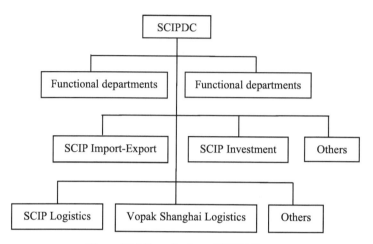

Figure 17.2 Organization of SCIPDC.

At the core of the complex is a 900,000-ton/yr cracker. It was built in 2005 at a cost of US$2.7 bn. The sponsors are BP, Sinopec, and Shanghai Petrochemical Corporation, and their equity holdings are 50, 30, and 20 per cent respectively. The joint-venture special purpose vehicle is SECCO.

The estimated capacities of downstream plants are all world scale, and comprise, among others,

- 600,000 ton/yr of polyethylene;
- 590,000 ton/yr of propylene;
- 500,000 ton/yr of aromatics;
- 500,000 ton/yr of styrene;
- 300,000 ton/yr of polystyrene;
- 260,000 ton/yr of acrylonitrile;
- 250,000 ton/yr of polypropylene; and
- 150,000 ton/yr of butadiene.

The nearly 30 km^2 world-class complex is supported by an impressive array of infrastructure including road, rail, and shipping. The total private and public investment in the complex in Phase I (2001–5), inclusive of land reclamation and other infrastructure, was around a massive US$18 bn of which nearly half were from the private sector.

17.4 Vopak Shanghai Logistics Company

For a closer look at how a petrochemical project is financed, we shall examine Vopak Shanghai Logistics Company (VSLC), a chemical storage joint venture between Royal Vopak and SCIPDC (see Figure 17.2).

Royal Vopak (hereafter "Vopak") is the world's largest independent tank terminal operator. It specializes in storing and handling liquid and gaseous chemical and oil products as well as complementary logistic services (such as drumming of liquid products) at its terminals.

In 2005, Vopak operated a network of 75 tank terminals with a total storage capacity in excess of 20 million m³. These terminals are strategically located along major shipping routes.

Vopak is organized by market regions using the ports of the world's largest refineries (Rotterdam-Antwerp, Houston, and Singapore) as hubs providing the full range of terminal services including storage, import, transshipment, and export. These terminals are integrated within a major petrochemical complex or refinery and upstream and downstream producers outsource the terminal (storage) function to Vopak. The company's strengths are its safety record, an experienced workforce, quality service, financial health, and track record in strategic cooperation with third parties.

By the late 1990s, Vopak could no longer ignore the strategic importance of setting up storage facilities in several of China's rapidly expanding network of petrochemical complexes to service its demand for petrochemical products. SCIP was a natural choice for Vopak to build its terminals given its proximity to the large Shanghai market.

As noted above, Vopak Shanghai Logistics Company (VSLC) is a 50:50 industrial chemical storage joint venture between Vopak Logistics Asia Pacific BV and SCIPDC. The project was initiated in 2001 and has an initial capacity of 240,000 m³ of storage in Phase I (completed in 2005), expandable to 700,000 m³ by 2010 in Phase II.

For a variety of reasons, the project scope for the VSLC joint venture was limited to jetty services, storage, and handling, and excluded land-based distribution. For Phase I, the project includes a jetty (six berths), warehouse, drumming lines, and rail connection. In all, the project occupies a total area of 50 ha.

17.5 Project finance structure

The financing structure was based on equity and limited-recourse debt. The equity share of each joint venture partner was US$40 m, and the debt for Phase I was US$160 m, making a total of US$240 m in investment. The 10-year debt, one of the first few project finance loans to be denominated in Yuan, was raised from a syndicate of international and Chinese lenders (with DBS bank as lead arranger) at an estimated fixed rate of 8 per cent, or a real rate of 7 per cent (although China's inflation rate of about one per cent for 2001 may not be reliable).

The debt in Yuan exposed Vopak to currency risk if the Yuan appreciated, but it can be hedged without much difficulty. Since the Yuan is likely to appreciate in the long term, the denomination of the debt in Yuan made sense from the lenders' perspective.

For Vopak, its equity share of US$40 m minimizes its exposure to the Chinese market by developing the project into two stages. Based on its annual reports, its current ratio of current assets to current liabilities was 1.1 in 2001 (1.52 in 2005 after divesting non-core assets) when the project was conceived, which meant Vopak needed to be reasonably liquid and should not invest too much equity into this project. Vopak's debt/equity ratio for the same year was 1090.1/339 (in million euros) = 3.22. The relatively high ratio reflects Vopak's aggressive expansion strategy and the high capital intensity of its chemical storage business. By 2005, its debt equity ratio had fallen to 0.72.

Vopak requires a rate of return on capital employed (ROCE) of 16 per cent (Vopak *Annual Report*, 2002). There are actually two definitions of "capital employed," namely,

ROCE = Net profit/Total investment; and
ROCE = Net profit/Equity shareholders' funds.

It can be seen that both definitions are similar to equity or project IRR discussed in Chapter 6 except that discounting is not used. Hence, ROCE is a simple indicator of profitability. In addition, Vopak adds about 2–3% as country risk premium to its projects in China.

17.6 Land option

Land for Phase I (25 ha) was secured for an estimated US$70 per m^2 ($S$, or current price) for a 50-year lease. The land price is estimated from prices of similar sites in the vicinity, and the lease period lies within Chinese guideline of between 40 to 70 years depending on the type of development. There was an option to lease the adjoining 25 ha for Phase II for an undisclosed sum with an expiry of eight years (T). Assuming a strike price (K) of US$80 per m^2 and volatility of land prices (σ) of 0.3, then

$S = 70$;　　　　　$K = 80$;　　　　　r = risk-free interest rate = 0.04;
$T = 8$ years;　　$\sigma = 0.3$.

Using the Black-Scholes option pricing model,

$$d_1 = \frac{\log(S/K)+(r+0.5\sigma^2)T}{\sigma\sqrt{T}} = \frac{\log(70/80)+[0.04+0.5(0.3^2)]8}{0.3\sqrt{8}}$$

$$= (-0.1335 + 0.68)/0.8485 = 0.644.$$

$$d_2 = d_1 - \sigma\sqrt{T} = 0.644 - 0.3\sqrt{8} = -0.205.$$

From the Appendix, the cumulative probabilities are

$$N(0.644) = 0.740; \text{ and}$$
$$N(-0.205) = 0.420.$$

Hence, the price of the call is

$$C = SN(d_1) - K^{-rT}N(d_2)$$
$$= 70(0.740) - 80e^{-0.04(8)}(0.420)$$
$$= 51.80 - 24.40 = \$27.40 \text{ per m}^2.$$

The call price is relatively high because of the assumptions of a relatively low strike price despite Shanghai's over-heated property market of the early 2000s, high volatility arising from more recent curbs to cool the property market, and relatively long term to expiry (eight years). For a market that has little "memory" of a down trend in property prices, volatility is difficult to estimate. Obviously, the property boom cannot last forever.

The call is likely to be underpriced to encourage Vopak to invest in Phase II should Phase I prove profitable.

17.7 Risk management

Vopak uses a high-level Enterprise Risk Management Framework (ERMF) adapted from the 1992 Committee of Sponsoring Organizations of the Treadway Committee (COSO) model for internal controls in accounting and the COSO ERM model (2002).

ERMF has eight components, namely,

- internal environment, which establishes an entity's risk culture;
- objective setting of risk tolerance;
- event identification;
- risk assessment;
- risk response;
- control activities;
- information and communication; and
- monitoring.

These components are similar to the risk management framework discussed in Chapter 9.

Vopak reduces its market risk through long-term (15 years) take or pay contracts with its customers prior to project completion. For the SCIP project, the first eleven customers were established international chemical firms such as BP Chemical, BASF, Huntsman, and Bayer.

In managing political risk, Vopak locates its hub at three politically stable countries, namely, Rotterdam-Antwerp, Houston, and Singapore. Shanghai is seen as a rapidly expanding market, and Vopak is likely to expand its operations there. In all, its assets and revenues are diversified across 29 countries.

At the operational level, Vopak has put in place a comprehensive in-house risk management system to mitigate operational risks that include safety, health, and environmental (SHE) risks. The latter is taken seriously since about 17% of its investments were in SHE activities (*Annual Report*, 2002). It also invested heavily in its information infrastructure to streamline commercial, operating, and business processes. In the SCIP project, Vopak exercises control over the commercial and operational matters where it has comparative advantage to effectively manage the risks.

Vopak's exposure to credit risk is relatively small since its customers are reputable oil and chemical producers. Moreover, since it provides primarily storage services, the value of chemicals stored is much greater than the fee it charges, and Vopak has the right of retention if unpaid.

In terms of financial risks, most of its loans are at fixed rates to cover interest rate fluctuations. This consideration is important for any capital-intensive firm. Where possible and at reasonable cost, Vopak hedges about half of its net currency balances.

Questions

1   Interestingly, Vopak has not been all that enthusiastic about oil terminals in China. What are the obstacles?

2   What are the strengths and weaknesses of Vopak's enterprise risk management system?

3   Explain why Vopak's project finance loans are likely to contain fewer covenants.

Appendix: Cumulative standard normal distribution

The table shows areas for

$$N(z) = \int_{-\infty}^{z} f(u)\,du$$

where $f(u)$ is the standard normal density function, e.g. $N(-1.21) = 0.113$, $N(1.33) = 0.908$.

z	0.00	0.01	0.02	0.03	0.04	0.05	0.06	0.07	0.08	0.09
−2.9	0.002	0.002	0.002	0.002	0.002	0.002	0.002	0.001	0.001	0.001
−2.8	0.003	0.002	0.002	0.002	0.002	0.002	0.002	0.002	0.002	0.002
−2.7	0.003	0.003	0.003	0.003	0.003	0.003	0.003	0.003	0.003	0.003
−2.6	0.005	0.005	0.004	0.004	0.004	0.004	0.004	0.004	0.004	0.004
−2.5	0.006	0.006	0.006	0.006	0.006	0.005	0.005	0.005	0.005	0.005
−2.4	0.008	0.008	0.008	0.008	0.007	0.007	0.007	0.007	0.007	0.006
−2.3	0.011	0.010	0.010	0.010	0.010	0.009	0.009	0.008	0.009	0.008
−2.2	0.014	0.014	0.013	0.013	0.013	0.012	0.012	0.012	0.011	0.011
−2.1	0.018	0.017	0.017	0.017	0.016	0.016	0.015	0.015	0.015	0.014
−2.0	0.023	0.022	0.022	0.021	0.021	0.020	0.020	0.019	0.019	0.018
−1.9	0.029	0.028	0.027	0.027	0.026	0.026	0.025	0.024	0.024	0.023
−1.8	0.036	0.035	0.034	0.034	0.033	0.032	0.031	0.031	0.030	0.029
−1.7	0.045	0.044	0.043	0.042	0.041	0.040	0.039	0.038	0.038	0.037
−1.6	0.055	0.054	0.053	0.052	0.050	0.049	0.048	0.047	0.046	0.045
−1.5	0.067	0.066	0.064	0.063	0.062	0.061	0.059	0.058	0.057	0.056
−1.4	0.081	0.079	0.078	0.076	0.075	0.074	0.072	0.071	0.069	0.068
−1.3	0.097	0.095	0.093	0.092	0.090	0.089	0.087	0.085	0.084	0.082
−1.2	0.115	0.113	0.111	0.109	0.107	0.106	0.104	0.102	0.100	0.098
−1.1	0.136	0.134	0.131	0.129	0.127	0.125	0.123	0.121	0.119	0.117
−1.0	0.159	0.156	0.154	0.152	0.149	0.147	0.145	0.143	0.140	0.138
−0.9	0.184	0.181	0.179	0.176	0.174	0.171	0.169	0.166	0.164	0.161
−0.8	0.212	0.209	0.206	0.203	0.200	0.198	0.195	0.192	0.189	0.187
−0.7	0.242	0.239	0.236	0.233	0.230	0.227	0.224	0.221	0.218	0.215
−0.6	0.274	0.271	0.268	0.264	0.261	0.258	0.255	0.251	0.248	0.245
−0.5	0.309	0.305	0.302	0.298	0.295	0.291	0.288	0.284	0.281	0.278
−0.4	0.345	0.341	0.337	0.334	0.330	0.326	0.323	0.319	0.316	0.312
−0.3	0.382	0.378	0.374	0.371	0.367	0.363	0.359	0.356	0.352	0.348
−0.2	0.421	0.417	0.413	0.409	0.405	0.401	0.397	0.394	0.390	0.386
−0.1	0.460	0.456	0.562	0.448	0.444	0.440	0.436	0.432	0.429	0.425
0.0	0.500	0.504	0.508	0.512	0.516	0.520	0.523	0.528	0.532	0.536
0.1	0.540	0.544	0.548	0.552	0.556	0.560	0.564	0.568	0.571	0.575
0.2	0.579	0.583	0.587	0.591	0.595	0.599	0.603	0.606	0.610	0.614
0.3	0.618	0.622	0.626	0.629	0.633	0.637	0.641	0.644	0.648	0.652
0.4	0.655	0.659	0.663	0.666	0.670	0.674	0.677	0.681	0.684	0.688
0.5	0.691	0.695	0.698	0.702	0.705	0.709	0.712	0.716	0.719	0.722
0.6	0.726	0.729	0.732	0.736	0.739	0.742	0.745	0.749	0.752	0.755

Appendix: Cumulative standard normal distribution (continued)

z	0.00	0.01	0.02	0.03	0.04	0.05	0.06	0.07	0.08	0.09
0.7	0.758	0.761	0.764	0.767	0.770	0.773	0.776	0.779	0.782	0.785
0.8	0.788	0.791	0.794	0.797	0.800	0.902	0.805	0.808	0.811	0.813
0.9	0.816	0.819	0.821	0.824	0.826	0.829	0.831	0.834	0.837	0.839
1.0	0.841	0.844	0.846	0.848	0.851	0.853	0.855	0.858	0.860	0.862
1.1	0.864	0.867	0.869	0.871	0.873	0.875	0.877	0.879	0.881	0.883
1.2	0.885	0.887	0.889	0.891	0.893	0.894	0.896	0.898	0.900	0.901
1.3	0.903	0.905	0.907	0.908	0.910	0.911	0.913	0.915	0.916	0.918
1.4	0.919	0.921	0.922	0.924	0.925	0.926	0.928	0.929	0.931	0.932
1.5	0.933	0.934	0.936	0.937	0.938	0.939	0.941	0.942	0.943	0.944
1.6	0.945	0.946	0.947	0.948	0.949	0.950	0.951	0.953	0.954	0.954
1.7	0.955	0.956	0.957	0.958	0.959	0.960	0.961	0.962	0.962	0.963
1.8	0.964	0.965	0.966	0.966	0.967	0.968	0.969	0.969	0.970	0.971
1.9	0.971	0.972	0.973	0.973	0.974	0.974	0.975	0.976	0.976	0.977
2.0	0.977	0.978	0.978	0.979	0.979	0.980	0.980	0.981	0.981	0.982
2.1	0.982	0.983	0.983	0.983	0.984	0.984	0.985	0.985	0.985	0.986
2.2	0.986	0.986	0.987	0.987	0.987	0.988	0.988	0.988	0.989	0.989
2.3	0.989	0.990	0.990	0.990	0.990	0.991	0.991	0.991	0.991	0.992
2.4	0.992	0.992	0.992	0.992	0.993	0.993	0.993	0.993	0.993	0.994
2.5	0.994	0.994	0.994	0.994	0.994	0.995	0.995	0.995	0.995	0.995
2.6	0.995	0.995	0.996	0.996	0.996	0.996	0.996	0.996	0.996	0.996
2.7	0.997	0.997	0.997	0.997	0.997	0.997	0.998	0.997	0.997	0.997
2.8	0.997	0.998	0.998	0.998	0.998	0.998	0.998	0.998	0.998	0.998
2.9	0.998	0.998	0.998	0.998	0.998	0.998	0.998	0.999	0.999	0.999
3.0	0.999	0.999	0.999	0.999	0.999	0.999	0.999	0.999	0.999	0.999

References

Alkerlof, G. and Yellen, J. (1986; Eds) *Efficiency wage models of the labor m*arket. London: Cambridge University Press.

Alonso, W. (1964) *Location and land use.* Massachusetts: Harvard University Press.

Amsden, A. (1992) *Asia's next giant.* New York. Oxford University Press.

Arrow, K. (1963) Uncertainty and the welfare economics of medical care. *American Economic Review,* 53, 941–69.

Arrow, K. and Debreu, G. (1954) Existence of an equilibrium for a competitive economy. *Econometrica,* 22, 265–90.

Asia Development Bank (2003) *Toward e-development in Asia and the Pacific.* Manila: ADB.

Atkin, M. and Glen, J. (1992) Comparing capital structures around the globe. *The International Executiv*e, 34, 369-87.

Averch, H. and Johnson, L. (1962) Behavior of firm under regulatory constraint. *American Economic Review,* 52, 1052–69.

Baumol, W. (1959) *Business behavior, value and growth.* New York: Macmillan.

Berle, A. and Means, G. (1932) *The modern corporation and private property.* New York: Commerce Clearing House.

Bernstein, L. (1996) The new religion of risk management. *Harvard Business Review,* March-April, 51–7.

Bird, R. and Vaillancourt, F. (1998, Eds) *Fiscal decentralization in developing countries.* New York: Cambridge University Press.

Bohi, D. (1981) *Analyzing demand behavior: a survey of energy elasticities.* Baltimore: Johns Hopkins Press.

Booth, L., Aivazian, V., Demirguc-Kunt, A. and Maksimovic, V. (2001) Capital structure in developing countries. *Journal of Finance,* 56, 87–130.

Borch, K. (1962) Equilibrium in a reinsurance market. *Econometrica,* 30, 424–44.

Borch, K. (1990) *Economics of insurance.* Amsterdam: North-Holland.

Borenstein, S. (2002) The trouble with electricity markets: understanding California's restructuring disaster. *Journal of Economic Perspectives,* 14, 191–21.

Borjas, G. (2004) *Labor economics.* New York: McGraw-Hill.

Bos, T. and Newbold, P. (1984) An empirical investigation of the possibility of stochastic systematic risk in the market model. *Journal of Business,* 57, 35–41.

Bossidy, L., Charan, R. and Burck, C. (2002) *Execution: The discipline of getting things done.* New York: Crown Publishing.

Box, G. and Jenkins, G. (1970) *Time series analysis.* San Francisco: Holden Day.

Brook, M. (2004) *Estimating and tendering for construction work.* Amsterdam: Elsevier.

Bruce, C. (1976) *Social cost benefit analysis.* Washington DC: World Bank.

Burt, D., Dobler, D. and Starling, S. (2003) *World class supply management.* New York: McGraw-Hill.

Chemmanur, T. and John, K. (1996) Optimal incorporation, structure of debt contracts, and limited recourse project financing. *Journal of Financial Intermediation,* 5, 372–408.

Chenery, H. and Bruno, M. (1962) Development alternatives in an open economy: the case of Israel. *Economic Journal*, 72, 79–103.

Clawson, M. and Knetsch, J. (1966) *Economics of outdoor recreation*. Baltimore: Johns Hopkins University Press.

Collins, J. and Porras, J. (1994) *Built to last*. New York: Collins Business.

Cummins, J. and Van Derhei, J. (1979) A note on the relative efficiency of property-liability insurance distribution systems. *Bell Journal of Economics*, 10(2), 709–19.

Cummins, D. and Weiss, M. (1993) Measuring cost efficiency in the property-liability insurance industry. *Journal of Banking and Finance*, 17, 463–81.

Dahl, R. (1961) *Who governs?* New Haven: Yale University Press.

Damodaran, A. (2001) *The dark side of valuation*. New York: Prentice Hall.

Davenport, P. (1995) *Construction claims*. Sydney: Federation Press.

Davis, J. (1994) The cross-section of realized stock returns: The pre-COMPUSTST evidence. *Journal of Finance*, 49, 1579–93.

Deal, T. and Kennedy, A. (1982) *Corporate cultures*. Harmondsworth: Penguin.

Diamond, D. (1991) Debt maturity and liquidity risk. *Quarterly Journal of Economics*, 106, 709–37.

Diamond, P. (1968) The opportunity cost of public investment: comment. *Quarterly Journal of Economics*, 82, 682–8.

Drucker, P. (1967) *The effective executive*. New York: Perennial Library.

Eichenwald, K. (2005) *Conspiracy of fools*. New York: Broadway Books.

Evans, J. and Dean, J. (2003) *Total quality: Management, organization, and strategy*. New York: Thomson.

Esty, B. (2004) *Modern project finance: a casebook*. New York: Wiley.

Fama, E. and French, K. (1992) The cross-section of expected stock returns. *Journal of Finance*, 47, 427–65.

Filippini, M. (1995) Electricity demand by time of use. *Energy Economics*, 17, 197–204.

Finnerty, J. (1996) *Project financing*. New York: Wiley.

Fox, L. (2003) *Enron: The rise and fall*. New York: Wiley.

Friedman, M. and Savage, L. (1948) The utility analysis of choices involving risk. *Journal of Political Economy*, 56, 279–304.

Gans, H. (1962) *The urban villagers*. Glencoe: The Free Press.

Gerstner, L. (2002) *Who says elephants can't dance?* London: HarperCollins.

Gillen, D. and Lall, A. (1997) Developing measures of airport productivity and performance: an application of data envelopment analysis. *Transportation Research A*, 33, 261–73.

Gramsci, A. (1971) *Selections from prison notebooks*. New York: International Publishers.

Griliches, Z. (1977) Estimating the returns to schooling: some econometric problems. *Econometrica*, 45(1), 1–22.

Grodinsky, J. (2000) *Transcontinental railway strategy, 1869-1893: A study of businessmen*. New York: Beard Books.

Halligan, P., Hester, W. and Thomas, H. (1987) Managing unforeseen site conditions. *Journal of Construction Engineering*, 113(2), 273–87.

Hamel, G. and Prahalad, C. (1994) *Competing for the future*. Boston: HBS Press.

Hammer, M. and Champy, J. (1993) *Reengineering the corporation*. New York: Harper-Business.

Hanweck, G. and Hogan, A. (1996) The structure of the property/casualty insurance industry. *Journal of Economics and Business*, 48, 141–55.

Harberger, A. (1972) *Project evaluation*. London: Macmillan.

Hart, O. and Moore, J. (1995) Debt and seniority: An analysis of hard claims in constraining management. *American Economic Review*, 85, 567–87.

Hekman, J. (1985) Rental price adjustment investment in the office market. *AREUEA Journal*, 13(1), 2–47.

Hellriegel, D., Jackson, S. and Slocum, J. (2005) *Management: a competency based approach*. Ohio: Thomson Learning.

Higgins, J. (1994) *The management challenge*. London: Macmillan.

Hoffman, S. (2001) *The law and business of international project finance*. New York: Kluwer Law International.

Hooper, P. and Hensher, D. (1997) Measuring total factor productivity of airports – an index number approach. *Transportation Research E*, 33, 249–59.

Hull, J. (2003) *Options, futures, and other derivatives*. New Jersey: Prentice Hall.

Humphrey, W. (1989) *Managing the software process*. New York: Addison-Wesley.

Hunter, F. (1953) *Community power structure: a study of decision makers*. Chapel Hill: University of North Carolina Press.

Imai, M. (1991) *Kaizen*. New York: McGraw-Hill.

Jenkins, G. (1997) Project analysis at the World Bank. *American Economic Review*, 87(2), 38–42.

Jensen, M. and Meckling, W. (1976) Theory of the firm: managerial behavior, agency costs, and ownership structure. Journal of Financial Economics, 3, 305–60.

Jones, C. and Williams, J. (1998) measuring the social return to R & D. *Quarterly Journal of Economics*, 113(4), 1119–35.

Jorion, P. (2002) *Value at risk*. New York: McGraw-Hill.

Joskow, P. (1973) Cartels, competition and regulation in the property-liability insurance industry. *Bell Journal of Economics*, 4(2), 375–427.

Kahneman, D. and Tversky, A. (1979) Prospect theory: an analysis of decision under risk. *Econometrica*, 47, 263–91.

Kaplan, R. and Norton, D. (1996) *The balanced scorecard*. Boston: HUP.

Kaplan, R. and Norton, D. (2001) *The strategy-focused organization*. Boston: HUP.

Kaplan, R. and Norton, D. (2004) *Strategy maps*. Boston: HUP.

Kettell, B. (2002) *Valuation of Internet and technology stocks*. London: Butterworth-Heinemann

Keynes, J. (1936) *The general theory of employment, interest, and money*. London: Macmillan.

Khan, M. and Parra, R. (2003) *Financing large projects*. Singapore: Pearson Prentice Hall.

Kirzner, I. (1985) *Discovery and the capitalist process*. Chicago: Chicago University Press.

Kleimeier, S. and Megginson, W. (2000) *An empirical analysis of limited recourse project finance*. Limburg Institute of Financial Economics Working Paper 01-03. Maastricht: LIFE.

Knight, F. (1921) *Risk, uncertainty, and profit*. Boston: Houghton Mifflin.

Kornai, J. (1986) The soft budget constraint. *Kyklos*, 39(1), 3–30.

Kouzes, J. and Posner, B. (1995) *The leadership challenge: How to keep getting extraordinary things done in organizations*. San Francisco: Jossey-Bass.

Krutilla, J. (1967) Conservation Reconsidered. *American Economic Review*, September, 777–86.

Landes, D. (1999) *The wealth and poverty of nations*. New York: W. W. Norton.

Lang, L. (1998) *Project finance in Asia*. New York: Elsevier.

Leontieff, W. (1986) *Input-output economics*. New York: Oxford University Press.

Lev, B. (2001) *Intangibles: management, measurement, and reporting*. Washington DC: Brookings Institute

Lewis, A. (1954) Economic development with unlimited supplies of labor. *Manchester School*, 22, 139–91.

Lewis, A. (1955) *The theory of economic growth*. Illinois: Irwing.

Lipton, M. (1977) *Why poor people stay people: urban bias in world development*. Mass.: Harvard University Press.

Little, I. and Mirrless, J. (1974) *Project appraisal and planning for developing countries*. London: Heinemann.

Little, I. and Mirrless, J. (1990) Project appraisal and planning twenty years on. In S. Fisher, D. de Tray and S. Shah (eds) *Proceedings of the World Bank Annual Conference on Development Economics*. Washington DC: World Bank.

Lucas, R. (1976) Econometric policy evaluation: a critique. *Carnegie-Rochester Conference Series on Public Policy*, 1, 19–46.

Luenberger, D. (1998*) Investment science*. New York: Oxford University Press.

Lutz, J., Hancher, D. and East, E. (1990) Framework for design-quality-review database system. *Journal of Management in Engineering*, 6(3), 296–312.

Machlup, F. (1946) Marginal analysis and empirical research. *American Economic Review*, 63, 519-54.

Manas, T. and Graham, M. (2003) *Creating a total rewards strategy*. New York: AMACOM.

Marglin, S. (1974) What do bosses do? In A. Gorz (ed.), *The Division of Labor*. Hassocks: Harvester.

Marris, R. (1964) *The economic theory of managerial capitalism*. Glencoe: Free Press.

Marx, K. (1963) *The Eighteenth Brumaire of Louis Bonaparte*. New York: International Publishers.

McDonald, J. (2002) A survey of econometric models of office markets. *Journal of Real Estate Literature*, 10(2), 223–42.

McLean, B. and Elkind, P. (2003) *The smartest guys in the room*. London: Penguin.

Miles, R. and Snow, C. (1994) *Fit, failure, and the hall of fame*. New York: Free Press.

Miliband, R. (1969) *The State in capitalist society*. New York: Basic Books.

Mintzberg, H. (1993) *Designing effective organizations*. New Jersey: Prentice Hall.

Mitchell, R. and Carson, R. (1989) *Using surveys to value public goods*. Washington DC: Resources for the Future.

Modigliani, F. and Miller, M. (1958) The cost of capital, corporate finance, and the theory of investment. *American Economic Review*, 48, 261–97.

Molotch, H. (1976) The city as a growth machine: toward a political economy of place. *American Journal of Sociology*, 82 (2), 309–32.

Myers, S. (1977) Determinants of corporate borrowings. *Journal of Financial Economics*, 5, 147–75.

Nadiri, I. (1993) *Innovations and technological spillovers*. NBER Working Paper No. 4423. Boston: NBER.

Nairn, A. (2002) *Engines that move markets*. New York: Wiley.

Netherton, R. (1983) *Construction contract claims: Causes and methods of settlement*. Washington DC: Transport Research Board.

Nevitt, P. and Fabozzi, F. (2000) *Project financing*. London: Euromoney Books.

Nurkse, R. (1953) *Problems of capital formation in underdeveloped countries*. Oxford: Blackwell.

O'Connor, J. (1973) *The fiscal crisis of the State*. New York: St Martin's Press.

Oates, W. (1972) *Fiscal federalism*. New York: HBJ.

Ofek, E. (1993) Capital structure and firm response to poor performance: An empirical analysis. *Journal of Financial Economics*, 34, 3–30.

Pahwa, H. (1991) *Project financing*. New Delhi: Bharat Law House.

Pande, P., Neuman, R. and Cavanagh, R. (2000) *The six-sigma way*. New York: McGraw-Hill.

Pickrell, D. (1989) *Urban rail transit projects*. Massachusetts: US Department of Transport.

Porter, M. (1980) *Competitive strategy*. New York: The Free Press.

Porter, M. (1985) *The competitive advantage*. New York: Free Press.

Poulantzas, N. (1969) The problem of the capitalist State. *New Left Review*, 58, 67–78.

Pratt, J. (1964) Risk aversion in the small and in the large. *Econometrica*, 32, 122–36.

Prest, A. and Turvey, R. (1965) Cost-benefit analysis: a survey. *Economic Journal*, 75, 683–735.

Rajan, R. and Zingales, L. (1995) What do we know about capital structure? Some evidence from international data. *Journal of Finance*, 50, 1421–60.

Roll, R. and Ross, S. (1994) On the cross-sectional relation between expected returns and betas. *Journal of Finance*, 49, 101–21.

Rosen, K. (1984) Toward a model of the office building model. *AREUEA Journal*, 12(3), 261-9.

Sanders, H. (2002) Convention myths and markets: a critical review of convention center feasibility studies. *Economic Development Quarterly*, 16(3), 195–210.

Sandmo, A. and Dreze, J. (1971) Discount rates for public investment in closed and open economies. *Economica*, 38, 395–412.

Schumpeter, J. (1942) *Capitalism, socialism and democracy*. New York: Harper.

Semple, C. Hartman, F. and Jergeas, G. (1994) Construction claims and disputes: Causes and time/cost overruns. *Journal of Construction Engineering and Management*, 120(4), 785–95.

Shackle, G. (1979) *Imagination and the nature of choice*. Edinburgh: Edinburgh University Press.

Shah, S. and Thakor, A. (1987) Optimal capital structure and project financing. *Journal of Economic Theory*, 42, 209–43.

Simon, H. (1957) *Models of man*. New York: Wiley.

Sims, C. (1980) Macroeconomics and reality. *Econometrica*, 48(1), 1–48.

Software Engineering Institute (1995) *Capability Maturity Model: guidelines for improving the software process*. New York: Addison-Wesley.

Software Engineering Institute (2006) *Risk management principles*. Pittsburgh: SEI. Retrieved 1 July 2006 from http://www.sei.cmu.edu/risk/principles.html.

Solow, R. (1997) *Learning from "learning by doing": lessons for economic growth*. Palo Alto: Stanford University Press.

Squire, L. (1989) Project evaluation in theory and practice. In H. Chenery and T. Srinivasan (eds) *Handbook of Development Economics II*. Amsterdam: North-Holland.

Stone, C. (1993) Urban regimes and the capacity to govern: a political economy approach. *Journal of Urban Affairs*, 15(1), 1–28.

Strong, R. (2005) *Derivatives*. Ohio: Thomson.

Tan, T. (2000) *e-Government: outlining the strategic thrusts*. Speech at opening ceremony of CommunicAsia 2000, Singapore.

Tan, W. (2004) *Practical research methods*. Singapore: Pearson Prentice Hall.

Tan, W. (2006) *Principles of housing investment*. Singapore: Pearson Prentice Hall.

Taylor, J. (1993) *Macroeconomic policy in a world economy*. New York: W. W. Norton.

Taylor, L. (1975) The demand for electricity: A survey. *Bell Journal of Economics*, Spring, 74–110.

Thorner, D. (1951) Great Britain and the development of India's railways. *Journal of Economic History*, 11(4), 389–402.

Tiebout, C. (1956) A pure theory of local expenditures. *Journal of Political Economy*, 64(5), 416–24.

Titman, S. and Wessels, R. (1988) The determinants of capital structure choice. *Journal of Finance*, 43, 1–19.

Tversky, A. and Kahneman, D. (1974) Judgment under uncertainty: heuristics and biases. *Science*, 211, 1124–30.

UNIDO (1972) *Guidelines for project evaluation*. New York: UN.

Vintner, G. (1998) *Project finance*. London: Sweet & Maxwell.

Von Neumann, J. and Morgenstern, O. (1944) *Theory of games and economic behavior*. New Jersey: Princeton University Press.

Wade, R. (1990) *Governing the market: Economic theory and the role of government in East Asian industrialization*. New Jersey: Princeton University Press.

Waring, A. and Glendon, I. (1998) *Managing risk*. London: Thomson Learning.

Watt, R. (2002) Defending expected utility theory. *Journal of Economic Perspectives*, 16, 227–9.

Williamson, O. (1975) *Markets and hierarchies*. New York: The Free Press.

Willig, R. (1976) Consumer's surplus without apology. *American Economic Review*, 66(4), 589–97.

Wilmott, P., Howison, S. and Dewynne, J. (1995) *The mathematics of financial derivatives*. New York: Cambridge University Press.

Womack, J. and Jones, D. (1998) *Lean thinking*. London: Touchstone Books.

World Bank (1993) *The East Asian miracle*. New York: Oxford University Press.

World Bank (1997) *The State in a changing world*. Washington DC: World Bank.

World Bank (2002) *Building institutions for markets*. Washington DC: World Bank.

Index